Biotechnology and Fungal Differentiation

Biotechnology and Fungal Differentiation

FEMS SYMPOSIUM NO. 4

Edited by

J. MEYRATH

Institut für angewandte Mikrobiologie
Vienna, Austria

J. D. BU'LOCK

Microbial Chemistry Laboratory,
The University, Manchester, England

1977

***Published by* ACADEMIC PRESS**
London · New York · San Francisco

on behalf of
THE FEDERATION OF EUROPEAN MICROBIOLOGICAL SOCIETIES

ACADEMIC PRESS INC. (LONDON) LTD.
24/28 Oval Road,
London NW1

United States Edition published by
ACADEMIC PRESS INC.
111 Fifth Avenue
New York, New York 10003

Library of Congress Catalog Card Number: 77-088849
ISBN 0-12-493550-8

Printed in Great Britain by
The Scolar Press Ltd.
Ilkley, Yorkshire

CONTRIBUTORS

ANDERSON, J. G.
Department of Applied Microbiology, Strathclyde University, Glasgow, Scotland.

AUNSTRUP, K.
Novo Industri A/S, DK-2880 Bagsvaerd, Denmark.

DAVIS, B.
Department of Applied Microbiology, Strathclyde University, Glasgow, Scotland.

DEANS, S. G.
Department of Applied Microbiology, Strathclyde University, Glasgow, Scotland.

DOBERSKY, P.
Department of Technical Microbiology, Institute of Microbiology, Czechoslovak Academy of Sciences, Prague 4, CSSR.

ESSER, K.
Lehrstuhl für Allgemeine Botanik, Ruhr-Universität, Postfach 102 148, 4630 Bochum, Germany.

IVANOVA, G. S.
Institute of Biochemistry and Physiology of Micro-organisms, USSR Academy of Sciences, Pushchino, Moscow region, USSR.

KATINGER, H. W.-D.
Institut für Angewandte Mikrobiologie, Universität für Bodenkultur, A-1190 Vienna, Austria.

MANACHERE, G.
Université Lyon I, Laboratoire de Mycologie au C.N.R.S. n° 44, Département de Biologie, 43, Bd du 11 Novembre 191, 69621 Villeurbanne, France.

MEINHARDT, F.
Lehrstuhl für Allgemeine Botanik, Ruhr-Universität, Postfach 102 148, 4630 Bochum, Germany.

NOVAK, M.
Department of Technical Microbiology, Institute of Microbiology, Czechoslovak Academy of Sciences, Prague 4, CSSR.
SHEPHERD, D.
Microbiology Section, Research Department, Nestle Products Technical Assistance Co. Ltd., CH-1350 Orbe, Switzerland.
SIKYTA, B.
Department of Technical Microbiology, Institute of Microbiology, Czechoslovak Academy of Sciences, Prague 4, CSSR.
SMITH, J. E.
Department of Applied Microbiology, Strathclyde University, Glasgow, Scotland.
STAHL, U.
Lehrstuhl für Allgemeine Botanik, Ruhr-Universität, Postfach 102 148, 4630 Bochum, Germany.
STANTON, W. R.
Universiti Malaya, Lembah Pantai, Kuala Lumpur, Malaysia.
SZAJER, Cz.
Institute of Chemistry and Agricultural Technology, College of Agriculture, 20-033 Lublin, Poland.
TARGONSKI, Z.
Institute of Chemistry and Agricultural Technology, College of Agriculture, 20-033 Lublin, Poland.
TURIAN, G.
Laboratoire de Microbiologie, Département de Biologie végétale, Université, 1211 Genève 4, Switzerland.

FOREWORD

In this time and age of rapid moving specialization it is good to see groups of specialists coming together for general stock-taking, assessing the advancement in the various disciplines, analyzing the importance of the findings for one's own project. While applied science is in principle not different from pure science in approach or required ability, the danger of the two streams getting too unrelated is always there. Biotechnology and pure microbiology are mutually interdependent. This was the spirit of the 4th FEMS Symposium held in Vienna.

Fungi were the main theme of the Symposium. These organisms are at a level of complexity such that they lend themselves particularly well for studies on differentiation. Fungi, furthermore, are one of the most important groups of organisms in biotechnological processes and technologists are aware of the complex transformations and differentiations of the cultivated organism taking place in a fermenter or cultivation chamber, be it for the production of chemicals or for biomass. Emphasis of this interdependence is found in several chapters, such as the need of the biotechnologist to "bring engineering, biochemical and microbiological techniques together as elements of novel art, as distinct from mere refinements of ancient art" (Stanton).

To cover the field of fungal differentiation and fungal technology in detail would require the writing of a voluminous treatise. Thanks to the authors' abilities and devotion a volume has been produced which in the field of fungal differentiation and related biotechnological processes represents a concise and advanced analysis of the subjects.

J. MEYRATH
J. D. BU'LOCK

ACKNOWLEDGEMENTS

The publication of this book has been greatly facilitated by a subsidy of the Austrian Ministry of Science and Research. The University of Agriculture permitted the use of their facilities to hold the Symposium.

Thanks to the authors' discipline in handing in their manuscripts at an early date, there was no delay in publication.

The hard work of the Symposium Secretariat, Mrs. Gudrun Bayer Dipl.-Ing., and Mr. Ludwig Berzaczy, was very much responsible for the success of the Symposium.

Painstaking work has been provided in the preparation of the "camera-ready" copy of this book by Mrs. Ursula Stangl and by Mr. Ludwig Berzaczy.

J. MEYRATH

CONTENTS

FUNGAL DIFFERENTIATION

G. Turian

INTRODUCTION

Differentiation can be considered as resulting from the differential expression of the genome under different conditions and at different times (Turian, 1969; Smith & Berry, 1974). It implies the selective activation and expression of sporogenic genes which have been shown to be controlled at several levels of integration.

1) Environmental-nutritional. Starvation for C and/or N sources is the most generally effective condition to induce spore formation in all genetically competent fungi studied. Other unfavorable growing conditions, desiccation, high temperature, low or high pH (chlamydospores) can also be effective. The morphological expression can differ with the treatments as in *Physarum*, where simple starvation induces sclerotium formation while in the presence of niacin and then light it leads to sporulation. A problem is to define the smallest amount of vegetative growth required before the organism becomes competent to produce reproductive bodies. Zoospores of Chytridiomycetes appear to be able to attain a minimum (minicycle) limited to the wall of the encysted zoospore directly generating a new zoospore.

2) Metabolic. At the initiation of the vegetative hyphal differentiation, there is establishment of a cytoplasmic polarization gradient to which the wall-synthesizing apparatus of the cell can respond. Respiration appears to be of major importance in the maintenance of this polarized growth as revealed in Mucorales and Aspergillales by the requirement for mitochondrial functioning in the hyphal alternative prevailing over yeast-like growth under aerobic conditions. In the maintenance of the vegetative, hyphal polarity, the apical vesicles containing wall precursors and hydrolyzing or syn-

thesizing enzymes could move up along an electrochemical gradient set up by oxygen through the hyphal cytoplasm. It has been cytochemically shown that the ultimate amitochondrial tip of vegetative hyphae is the site of a high reducing power maintained as long as glycolysis is enforced there by catabolite repression (Turian, 1976a). From the tip down to the oxidative mitochondria-rich subapical zone, this establishes a redox gradient which could settle the electrochemical gradient (apex positive - less negative - compared to the subapical zone). Catabolite derepression following depletion of exogenous glucose leads to the disappearance of the vegetative polarization gradient when the mitochondria can invade the hyphal tip and bring there, at least in the 1st phase of sporogenesis, the metabolic energy required for the new sporal syntheses.

3) Informational-macromolecular. Sporogenic differentiation of *Dictyostelium discoideum* is accompanied by a net loss of RNA and protein, with a high rate of RNA turn-over contributing to a remodelling of the sporal ribosomes. During spherulation of *Physarum* there is a sharp decrease of the total amounts of RNA, proteins and DNA. This decrease of some species of RNA is only partially compensated by the appearance of new species. At sporulation, a net decrease of all these macromolecules also occurs. The light requirement is associated with some further DNA synthesis for mitosis and the translation of new proteins from the sporulating genome. Thus there are, in *Physarum*, both transcriptional control with selective activation of sporal genes, as shown by hybridization, and translational control over the synthesis of sporal proteins. In yeasts induced to sporulate, there is also a decrease in the level of RNA and ribosomes in the developing ascus; this contains less proteins but twice as much nuclear DNA as the vegetative cell. However, specific syntheses of both RNA (tRNA and probably mRNA) and proteins can be detected with isotopically labelled precursors at the sporulating phase. In filamentous fungi, all systems - from *Achlya* to *Trichoderma* - have also shown a decrease of RNA and proteins at the induction of sporulation followed by new syntheses reflecting a change in RNA transcription and a probably controlled translation of speci-

fic proteins. We are thus led to think that the common denominator of all these antigrowth-sporogenic inductions is the destruction of vegetative RNA and protein (with some extracellular release) following the switch-out of the vegetative DNA information and leading to the switch-on of the sporal genes and the resumal of more specific, sporal RNA and protein syntheses. However, and most significant nuclear DNA resists to this drastic macromolecular turn-over and tends even to increase to insure the divisions of nuclei required to populate the spores of Physarum, the ascospores of yeasts or the conidia of Neurospora and other molds. To deepen our understanding of these control devices, we have selected the two fungal systems that we know best, namely Allomyces and Neurospora.

THE ALLOMYCES SYSTEM

In these blastocladiaceous fungi, motile cells (zygotes, zoospores) germinating in dilute saline solution never differentiate further than the monopolar stage (Turian, 1958). Such an extreme premature monosporogenesis, also observed in Blastocladiella when its zoospores are germinated in dilute phosphate buffer, has been termed lag phase (Hennessy & Cantino, 1972) or minicycle when liberating 2 zoospores (Barstow & Lovett, 1975). When induced in the additional presence of traces of cycloheximide (actidione), a few mini-resistant sporangia could be induced in Allomyces (Turian, 1975). In the dilute saline solution, the transition from monopolar (rhizoidal) to bipolar (rhizoidal plus hyphal) stage only occurred in the presence of traces of amino-acids, especially glutamate, as first suggested by the similar morphogenic effect of contaminating bacteria presumably liberating amino-acids (Turian, 1958).

Bipolar growth and relatively extensive development of hyphae occurs on Machlis' ammonium-glucose medium. Amino-acids are therefore produced by reductive amination, a process sensitive to decoupling agents. As expected, 2, 4-dinitrophenol (above 10^{-4} M) could completely prevent hyphal growth on sporophytic plantlets remaining monopolarly (some times bipolarly) rhizoidal (Turian, 1958; 1964). Plantlets germinated in the ammonium medium in the presence of boric acid are even apolarly polyrhizoidal but most plantlets regain bipolar growth with slowly elongating hyphae in the additional presence of 10^{-3} M glutamate (Turian, 1958; 1974a). Similar monopolar development with single spores liberated from encysted zoospores has been obtained in Allomyces germinated in N-deficient medium and phosphate buffer (Youatt, 1976a).

For full expression of the type of "organ" differentiation (male and female gametangia, zoosporangia and resistant sporangia), a minimum of vegetative, hyphal growth appears to be required. Such cases of so-called premature differentiation have been obtained from gametes liberated in a dilute yeast extract-boric acid solution. Parthenogenetically germinating female gametes developed into gametangial couple(s) and the few zygotes into bunches of zoosporangia (Fig. 10 in Turian, 1964). The conjunction of an early starvation for a few key compounds, presumably organo-nitrogenous, in the dilute yeast extract and of the toxic action of the boric acid, presumably at the hyphal apex level, could have been the acting process in that premature differentiation. It has recently been found that shift-up of meiospores to a sublethal temperature (37°C for Allomyces) during their early germination in a nutrient medium similarly and more simply turned the vegetative apex to premature gametangial differentiation (Turian & Ojha, in preparation; Fig. 1). Such a standard technique is being applied to the study of selective inhibition of gametangia of one or other sex: the most interesting result obtained to date is the prevention by hydroxyurea (HU) of differentiation of males in a high proportion of gametangial couples. This might be a consequence of the blocking effect of HU on DNA synthesis which is more crucial for male than for female differentiation because of its necessary associated nuclear multiplication (Turian, 1961).

Actinomycin D, inhibitor of gene transcription by complexing DNA, has shown a controversial effect, exercising an anti-male effect in the actively growing A. arbuscula (Turian & Viswanathan, 1966) while only slowing down the subapical gametangium differentiation indiscriminately be it male (hypogynous A. arbuscula) or female

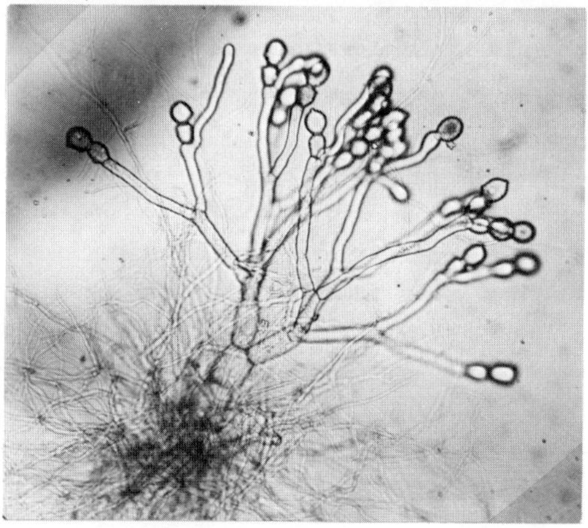

Fig. 1. Premature differentiation of gametangial couples induced by 20 → 37°C growth periods of *Allomyces arbuscula* (with M. Ojha).

(epigynous *A. macrogynus*) when applied to gametophytic cultures at their gametangial induction time (Youatt, 1976b). Actinomycin D added to shifted-up plantlets of both species of *Allomyces* has provided some evidence of prevention of male differentiation at high concentrations (20 μg/ml) only, the single gametangia left showing internally the wide diameter lipid crowns characteristic of female pregametic units (unpublished observations). When added at the induction time, actinomycin prevented ribosomal RNA synthesis occurring in the early phase of gametogenesis in *A. arbuscula* (Stumm & Croes, 1975).

Differentiation in the aquatic fungi involves rearrangement or disappearance of preexisting structures and general protein turnover has been shown for them (Murphy & Lovett, 1966; Burke, Seale & McCarthy, 1972; Timberlake, McDowell, Cheney & Griffin, 1973; Lodi & Sonneborn, 1974; Lovett, 1975). Losses of 25 % of the total intracellular protein and 35 % of the total RNA have been reported by Murphy & Lovett (1966) during sporulation of *Blastocladiella emersonii* and Lodi & Sonneborn (1974) have reported the release of a specific proteolytic enzyme during the later phase of differentiation in the same fungus at 150 - 250 min after induction.

The water mould Allomyces was induced to differentiate in a starvation salt solution (Machlis & Ossia, 1953; Kobr & Turian, 1967) which was then systematically scanned for UV-absorbing material. A correlation between release of a nucleic acid protein complex into the external medium and sporangial differentiation was found. Conditions preventing differentiation also stopped the release. Lethal lysis is not involved in this release process. Extracellular nucleic acids are very heterogenous, consisting of nucleotides as well as acid-precipitable nucleic acids. RNA was found to be associated with proteins and a hetero-duplex RNA-DNA associated with this nucleoprotein (Khandjian & Turian, 1976).

Release of many factors such as enzymes, antibiotics and nucleic acids reported as degradation products has been detected in the sporulation medium of various microorganisms and fungi (Murphy & Lovett, 1966; Mandelstam, 1971; Doi, 1972; Lodi & Sonneborn, 1974, references in Khandjian & Turian, 1976), but to our knowledge the release of nucleoprotein in the surrounding medium during differentiation has never been reported.

THE NEUROSPORA SYSTEM

A biconidial species of Neurospora such as N. crassa can differentiate two types of conidia, multinucleate macroconidia formed by blasto-arthroconidiogenesis and uninucleate microconidia generated by phialoconidiogenesis. Experimental situations have been devised which first limit Neurospora to the production of vegetative hyphae (submerged growth in media supplemented with Tween 80 or use of ammonium media) before conidiation is induced either by phosphate buffer replacement technique (Bianchi & Turian, 1967) or by transfer into lighted and properly aerated phosphate buffer (Stine, 1967). First generated are aerial hyphae (aerialogenesis, Nelson et al., 1975), most of which can then function as macro-conidiophores ("fluffy" mutants are blocked at this transition) which appear as more narrow in girth than other hyphae. In a next step, the tips of conidiophores, enriched in nuclei, swell and soon bud apically (Turian, Oulevey & Cortat, 1973). This bud enlarges into the first proconidium (blastoconidium) which in turn buds other presumptive macroconidia. The final step in the conidiogenetic process is the septation which will finally separate and disarticulate adjacent conidia then ontogenetically considered as blasto-arthroconidia (Turian, Oulevey & Cortat, 1973). Fact of functional significance, the initial nuclei have divided and their daughter nuclei have migrated into the elongating proconidial chain.

All vegetative hyphae appear capable of forming aerials. Thus, not only relatively old hyphae (at least 20 generations) but even hyphal outgrowths from newly germinated conidia can be induced. In the last case, the inducer of the so-called macroconidiogenic microcycle is heat (46°C during 15 h, Cortat & Turian, 1974); pretreatment with sublethal doses of metabolic inhibitors such as NaF (Timberlake & Turian, 1975) and 2,4-dinitrophenol (as suggested by Smith et al., 1976) can also be effective as can also be overcrowding of the germinating conidia (interaction of self-inhibitors, original observations by Ortega-Perez). Most drastic conditions combining effects of the heat-treatment (48 h at 46°C) and transfer on phosphate-buffered plates (pH 6.4) have led microconidia from both the amacroconidiate ("fluffy") mutant and from wild type selectively stimulated by iodoacetate in nitrate medium to produce microcyclic phialoconidiogenesis (Rossier, Ton-That & Turian, 1977).

Conidial induction is not only affected by the type and concentration of the carbon source (acetate versus sucrose-glucose) but also by the type of nitrogen source. Thus, in Neurospora, N provided as ammonium-citrate in a sucrose synthetic medium tends to favour vegetative hyphal growth with a minimum of conidia formed contrarily to the conidia-promoting effect of nitrate (Turian, 1964). The ammonium ion inhibition of conidiation also occurs in other fungi as shown in Aspergillus niger (Galbraith & Smith, 1969). In the presence of excess ammonium ions, conidiation does not occur in spite of glucose exhaustion but nitrate has no such inhibitory effect. Acetate, glyoxylate and, especially, glycine overcome ammonium ion inhibition in Neurospora; glyoxylate and most amino acids exert the same effect in Aspergillus.

By relying on the facts that sulphydryl groups have

been shown cytochemically to be most abundant in the vegetative hyphal tips (Zalokar, 1965) and therefore considered as functionally responsible for the high reducing power demonstrable in this apical region (Turian, 1976a), where it is also cytochemically thought to be in causal relationship with a high rate of alcoholic glycolysis involving the functioning of SH-enzymes such as hexokinase, glyceraldehyde dehydrogenase and alcohol dehydrogenase, it was expected that sulphydryl reagents would slow down the glycolytic activity considered as "vegetalizing" (Turian, 1975) and that switching to oxidative metabolism would induce conidiation, at least in Neurospora. That is what occurred when we added (Oulevey-Matikiam & Turian, 1968; Turian, 1974b) to vegetative mycelium growing in a proglycolytic medium (ammonium-citrate, high sugar) anti-sulphydryls such as the alkylating iodoacetate, an organo-mercuric compound (p-C1Hg-benzoate) or a soluble quinone (phthiocol - hydroxynaphthoquinone). That this last type of compound was effective at a low concentration incited us to consider the possibility of the progressive self-production of some type of quinone or quinonoid compound in developing hyphae which thus could get conidio-induced when a critical, anti-SH concentration was reached in the apices. Thinking that such a type of morpho-inducer would be accumulated in conidiogenetic (C) nitrate-grown cultures (C-medium, Turian, 1964), we attempted to extract it with acetone from highly conidiated mycelia (4 days). The yellow acetone extracts were further evaporated to dryness and soluble components re-extracted with sterile, slightly alkaline water. In preliminary attempts, when we added a few drops of our "yellowish watery extracts" to vegetative mycelia (M) pregrown (36-48 h) on our ammonium, M-medium and sterilely transfered on to glass-beads in Petri dishes containing only phosphate-buffer solution (pH 6.8), their surface quickly became felt-like (conidiophores) and more generally conidiated than controls (unpublished results, 1972). During these trials, we observed an unexpected intense production of aerial hyphae, abundantly conidiated, in the presence of traces of rifamycin + rifampicin (100 μg/ml, conferring a slightly yellow-orange staining to the buffer). This compound having a quinonic nucleus, the possibility still exists that its conidiogenic effect depends on its presumed anti-SH properties parallel to its well-known inhibiting action on mitochondrial RNA polymerase. In this relationship, it may be interesting to mention that interaction with both sulphydryl groups and mitochondrial respiration has been implicated in the sporal maturation and dormancy effect of a natural sulphydryl reagent isolated from sporulating gill tissue of the mushroom, Agaricus bisporus, and identified as a glutamyl-derivative of benzoquinone (Weaver, Rajagopalan, Handler, Jeffs, Byrne & Rosenthal, 1970).

Apices can also be regarded as the most photosensitive zone and as such, certain types of irradiation could modulate the level of metabolic activity and switch it to a differentiative orientation.

The best known case is that of induction of sporulation in Trichoderma viride (phialoconidiation) by blue light (Galun & Gressel, 1967). In Botrytis cinerea the blue light inhibitory effect on sporal induction is released by near UV irradiation (Tan & Epton, 1970); a rather similar antagonistic dual photosporogenetic effect mediated by "mycochrome" is described in species of Alternaria (Japanese workers). In Neurospora, conidiation can occur in the dark. However, some light as provided by luminescent tubes activates it. We have hypothesized that near-UV might be the acting component of this conidiogenic effect through its ability to antagonize the reducing apical activities. Through a system of mycelial growth in wide vertical tubes in the ammonium-high sugar (4 % sucrose) vegetalizing M medium, we could impose a vertical growth of parallel aerial hyphae maintainable either in full darkness or submitted to short exposures to a source of near-UV light (200 nm). When the apices received the optimum exposure of 3 minutes UV and were further incubated for 12 h at 25°C, they produced their first significantly countable apical macroconidia from the 8th hour of further incubation in full darkness while a first significant bunch of conidia appeared only from the 12th hour on the hyphal tips of the control tubes (unpublished observations with Harrington, 1974).

As preliminary conclusions here, we think that near-UV light has someway depressed the levels of NADH and -SH groups generated and implicated in the glycolytic activity enforced in the apices by the ammonium high sugar regime. The UV-shock has thus led to an earlier functioning of oxidative, conidiogenic conditions in these hyphal tips.

It is still hard to explain how Neurospora receives environmental signals and converts them to internal biochemical conditions which result in either the maintenance of mycelial growth or the induction of conidiation. By treatment with a glycolytic inhibitor, NaF (anti-enolase, etc.), a first answer could be brought to this question by separating induction from the actual process of conidiation and by analyzing physiological events occurring at the time of induction (Timberlake & Turian, 1975). The transient treatment of pregerminated macroconidia of N. crassa with NaF induced young, submerged cultures to prematurely differentiate macroconidia. This inductive treatment decreased the rate of respiration (with lower R.Q.) and reduced the relative concentration of nucleotide triphosphates, especially ATP. The NaF treatment sharply inhibited leucine incorporation into protein and adenosine incorporation into RNA. Most interesting, precursor incorporation into DNA was proportionally much less severely reduced. It therefore appears that at conidial and sporal induction in general, an inhibition of RNA and protein synthesis with a consequent lowering of their intracellular concentrations is a normal event in the necessary turnover of these macromolecules while some further synthesis of DNA is a required early event of the differentiation process. Preliminary analyses reveal that this is also what happens in the early phase of the heat-induced microcyclic macroconidiation.

The fact that following pre-germinative treatment of conidia by NaF, the prematurely induced conidiation occurred against a sharp drop in the RNA-protein synthesis compared to that of DNA suggested that normal conidiation is inversely paralleled by an important increase in the DNA/RNA ratio. This has recently been confirmed by comparing the development evolution of this ratio in mycelial maintained cultures on our ammonium-high (2 %) sucrose M-medium with those induced to rapid conidiation on the same but low-sugared (0.2 %) medium (Turian & Grange, 1976). It is of interest

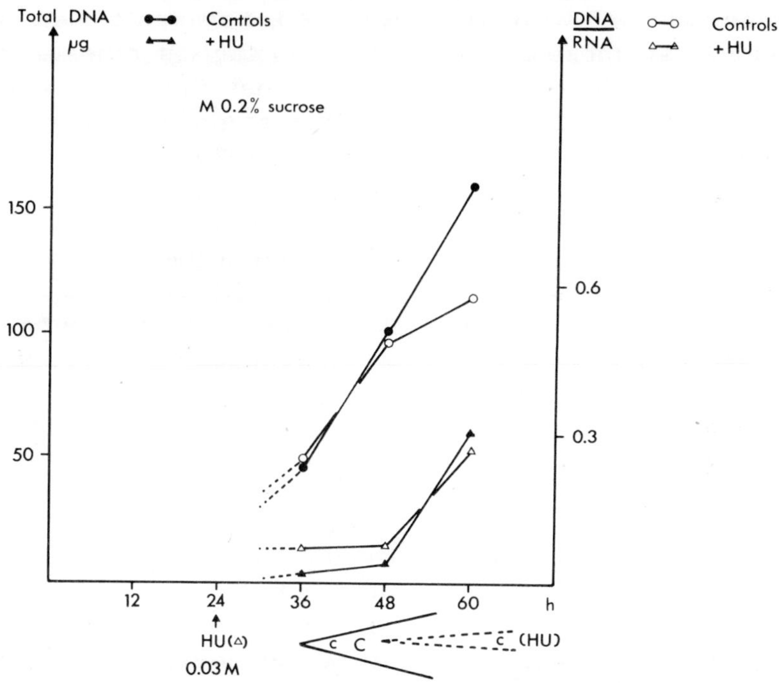

Fig. 2. Correlation between DNA per culture, DNA/RNA ratios, and amount of conidiation (low = c, dense = C) in *Neurospora crassa* grown in ammonium-low sucrose (0.2 %) medium in the absence or presence (arrow) of hydroxyurea (HU). Results from Turian & Grange (1976) and unpublished data.

to see (Fig. 2) that if DNA synthesis noticeably increases with the ontogenesis of the first conidia on aerial conidiophores (from the 36th hour on stationary cultures at 25°C), the decrease of total RNA only intervenes when there is a full cover of conidia (on 48 h cultures). This trend is also that of total proteins. It should mean that some new syntheses of RNA and proteins accompany the early conidial induction which for a short while compensate the previous sharp decrease which expresses the degradation of both vegetative RNA and protein from the declining vegetative state (sustaining light mycelium). In *Neurospora*, therefore, at the onset of differentiation, there seems to be an important macromolecular turnover with a minimum needed amount of new macromolecular syntheses. The need for an early synthesis of RNA for normal conidiation of Neurospora has first been shown by the inhibition of both aerialogenesis and subsequent conidiogenesis by 5-fluorouracil (Turian, 1970) as previously shown for conidiation in Trichoderma (Galun & Gressel, 1967). A significant increase in the cellular concentration of RNA has then been directly detected at the onset of aerialogenesis by Nelson, Selitrennikoff & Siegel (1975). Transcription for m-RNA sequentially preceeding conidiophore and conidia differentiations has been revealed by actinomycin inhibitions (Totten & Howe, 1971). When conidiogenic committment is reached, differentiation is unaffected by the drug (Turian & Bianchi, 1972).

There is also some novel, conidial type, protein synthesis as confirmed by the selective prevention of conidiogenesis by cycloheximide or by ethionine (Urey, 1971). This protein synthesis is paralleled by increases in the specific activities of several enzymes functioning in the glyoxylate and Krebs cycles as well as NADase (Combépine & Turian, 1970; Urey, 1971; for more details see Turian, 1974b; Schmit & Brody, 1976).

The fact that the elongating proconidiogenic chains were found to be densely populated by nuclei resulting from active mitotic activity (Turian, Oulevey & Cortat, 1973) suggested the need for new DNA synthesis associated with the induction process. This has been found to be so, and in a continuous process up to full conidiation, as confirmed by the increased DNA/RNA ratios measured in conidiating cultures (Turian & Grange, 1976). As expected, both synthesis and increased ratios could be prevented in the presence of hydroxyurea, with parallel inhibition of conidiogenesis, in N. crassa (Turian & Grange, 1976). This finding also applies to microcyclic macroconidiation with shortening of the conidial chains (progressing research with F. Grange). Other antimitotic inhibitors blocking DNA synthesis, such as 5-fluorodeoxyuridine (Nelson, Selitrennikoff & Siegel, 1975) or amethopterin (recent observations with M. Michéa), are also highly effective in preventing conidiogenesis which is also inactivated with UV irradiation (short waves, contrarily to the near UV effects mentioned above) delivered at the onset of induction (Nelson, Selitrennikoff & Siegel, 1975). This dependence upon DNA synthesis and the continuation of the mitotic cycles relates conidiogenesis in Neurospora to Wessells' model (1968) concerned with the cases of cell differentiation requiring DNA synthesis. As a corollary to this statement, parallely reached by Siegel's group (Nelson, Selitrennikoff & Siegel, 1975), we can now think that the heat microcyclic-inducing treatment (46°C) someway produces such an imbalance in favor of DNA synthesis resumed in the shifted-down (25°C) germinating conidia that all vegetative-type syntheses are prematurely deviated to conidial-type products.

REFERENCES

Barstow, W. E. & Lovett, J. S. (1975). Mycologia 67, 518-529.

Bianchi, D. E. & Turian, G. (1967). Experientia 23, 192.

Burke, D. J., Seale, T. W. & McCarthy, B. J. (1972). J. of Bacteriol. 110, 1055-1072.

Combépine, G. & Smith, J. E. (1969). J. Gen. Microbiol. 59, 31-45.

Cortat, M. & Turian, G. (1974). Arch. Microbiol. 95, 305-309.

Galbraith, J. & Smith, J. E. (1969). J. Gen. Microbiol. 59, 31-45.

Galun, F. & Gressel, J. (1966). Science 151, 696-698.

Hennessy, S. W. & Cantino, E. C. (1972). Mycologia 64, 1066-1087.

Khandjian, E. W. & Turian, G. (1076). Cell Differentiation 5, 171-188.

Kobr, M. & Turian, G. (1967). Arch. Mikrobiol. 57, 271-279.

Lodi, W. R. & Sonneborn, D. R. (1974). J. Bacteriol. 117, 1035-1042.

Lovett, J. S. (1975). Bacteriol. Rev. 39, 345-404.

Machlis, L. & Ossia, E. (1953). Amer. J. Bot. 40, 358-365.

Murphy, M. N. & Lovett, J. S. (1966). Develop. Biol. 14, 68-95.

Nelson, R. E., Selitrennikoff, C. P. & Siegel, R. W. (1975). Results and Problems in Cell Differentiation Vol. 7, 291-310.
Oulevey-Matikian, N. & Turian, G. (1968). Arch. Mikrobiol. 60, 35-58.
Rossier, C., Ton-That, C. T. & Turian, G. (1977). Exper. Mycology (in press).
Stine, G. J. & Clark, A. M. (1967). Canad. J. Microbiol. 13, 447-453.
Smith, J. E. & Galbraith, J. C. (1971). Advances in Microbial Physiology 5, 45-134.
Smith, J. E. & Berry, D. R. (1974). An introduction to Biochemistry of Fungal Development. 326 pp. Acad. Press, New York.
Smith, J. E., Gull, K., Anderson, J. G. & Deans, S. G. (1976). In "The Fungal Spore" Form and Function. (Edt.) Weber, D. J. & Hess, W. M., Wiley & Sons, New York.
Schmit, J. C. & Brody, S. (1976). Advances in Microbial Physiology 5, 45-134.
Stumm, C. & Croes, A. F. (1975). Arch. Microbiol. 102, 117-122.
Tan, K. K. & Epton, H. S. A. (1974). Trans. brit. mycol. Soc. 62, 105-112.
Timberlake, W. E., McDowell, L., Cheney, J. & Griffin, D. H. (1973). J. Bact. 116, 67-73.
Timberlake, W. E. & Turian, G. (1975). Arch. Microbiol. 104, 151-154.
Totten, R. E. & Howe, B. (1971). Biochem. Genet. 5, 521-532.
Turian, G. (1958). Rev. Cytol. Biol. vég. (Paris) 19, 241-272.
Turian, G. (1961). The Nucleus (Calcutta) 4, 151-156.
Turian, G. (1964). Bul. de la soc. bot. suisse 74, 242-257.
Turian, G. (1969). Différenciation Fongique. 144 pp. Paris, Masson & Cie.
Turian, G. (1970). Physiol. vég. 8, 375-386.
Turian, G. (1974a). Bull. Soc. Linn. Lyon, Vol. spécial. Travaux mycol. dédiés à R. Kühner.
Turian, G. (1974b). Ann. Rev. Phytopathol. 12, 129-137.

Turian, G. (1975). Trans. Brit. Mycol. Soc. 64, 367-380.

Turian, G. (1976a). Experientia 32, 989-991.

Turian, G. (1976b). In "The Fungal Spore", Chapt. 16. (Edt.). Weber, D. J. & Hess, W. M. Wiley & Sons, New York.

Turian, G. & Viswanathan, M. A. (1966). Pathol. Microbiol. 29, 705-729.

Turian, G. & Bianchi, D. E. (1972). Bot. Rev. 38, 119-154.

Turian, G., Oulevey, N. & Cortat, M. (1973). Ann. Microbiol. (Inst. Past.), 124, 443-458.

Turian, G. & Grange, F. (1976). C. R. Acad. Sc. Paris 282, 169-171.

Urey, J. C. (1971). Dev. Biol. 26, 17-27.

Weaver, R. F., Rajagopalan, K. V., Handler, P., Jefs, P., Byrne, W. L. & Rosenthal, D. (1970). Proc. Nat. Acad. Sci. (U.S.) 67, 1050-1056.

Wessells, N. K. (1968). Problems in the analysis of determination, mitosis and differentiation. In: Fleischmajer, R., Billingham, R. E. (Eds.) : Epithelial-Mesenchymal interactions, pp. 132-151. Baltimore, Md. Williams and Wilkins 1968.

Youatt, J. (1976a). Trans. Br. mycol. Soc. 67, 159-161.

Youatt, J. (1976b). Arch. Microbiol. 109, 163-166.

Zalokar, M. (1959). Am. J. Bot. 46, 555-559.

THE NATURE OF FUNGAL SPORULATION

J. E. Smith, S. G. Deans, J. G. Anderson
and B. Davis

INTRODUCTION

A spore is normally delimited from the thallus and is characterised by a minimal metabolic turnover, low water content and generally lacks cytoplasmic movement. Since spores normally occur at the end of the growth cycle when the parent mycelium is experiencing conditions of stress spores can be considered as a means of propagation, survival and dispersal. Fungal spores vary widely in the manner of their ontogeny and biological functions. In particular, within the truly filamentous fungi there exist motile zoospores, sessile sporangiospores, oidia, conidia, spermatia, aecidiospores, urediospores, teleutospores and chlamydospores (Weber & Hess, 1976).

The complex process of sporulation must be viewed as the integration of several biological events. This ability to change from a vegetative filamentous growth form to a reproductive and generally spherical spore form is the end result of a genetic competence responding to specific physicochemical environmental changes which in turn exerts the internal stimulus to regulate cellular activity. The consequence of this interplay is the establishment of a new developmental pattern with its characteristic metabolism and morphology (Turian, 1974; Smith, 1977). There have been several authoritative reviews on sporulation in filamentous fungi and these should be consulted for a deeper appreciation of this complex developmental process (Smith & Galbraith, 1971; Turian & Bianchi, 1972; Smith & Anderson, 1973; Smith & Berry, 1974; Smith, 1977; Lovett, 1975; Turian, 1974, 1975, 1976).

THE ONTOGENY OF FUNGAL SPORES

Asexual spores of filamentous fungi are formed mainly by fragmentation, fission, extrusion and cleavage (Carmichael, 1971). In fragmentation the cytoplasm becomes concentrated into particular cells in a filament leaving the remaining cells empty of cytoplasmic contents. Spore formation by fission occurs when cells within the mycelium break apart at double-walled septa while formation by extrusion occurs when the spores are produced as extensions from the ends or the sides of the sporogenous cells. Formation by cleavage occurs when the protoplasmic material of a cell splits or divides into fragments each of which becomes surrounded by a wall.

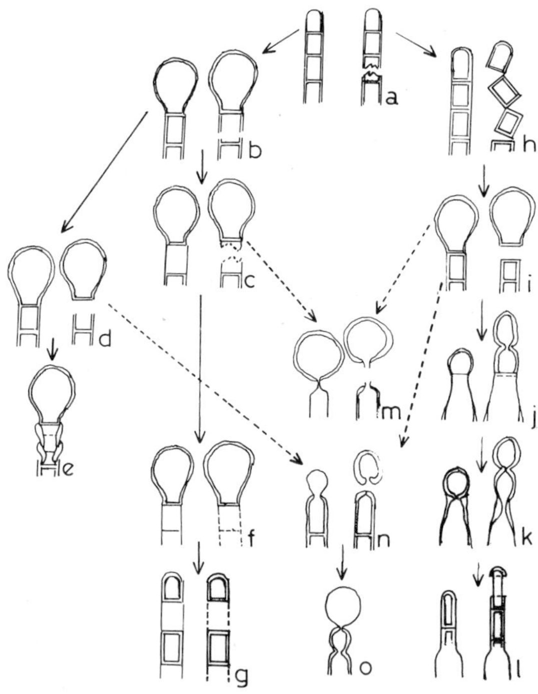

Fig. 1. Conidia classified by dehiscence; a, an undifferentiated hyphae releasing a propagule by fracture; b – g, conidia released by sacrifice of a supporting cell; h – l, conidia released by fission of a double septum; m – o, conidia released by fracture of a fine connective (from Carmichael, 1971).

The methods used to ultimately release the spores are shown in Fig. 1. Endogenous spores are released by the breakdown of the sporangium wall whereas chlamydospores do not possess any positive mechanism of release but rather rely on accidental mechanical breakage or digestion of surrounding mycelium by other organisms (see Kendrick, 1971).

Because of the bewildering range of form and function of asexual spores in the filamentous fungi the remainder of this chapter will be concerned with the mechanisms displayed in the Fungi Imperfecti and in particular with the genus *Aspergillus*.

The characteristic spore of the Fungi Imperfecti is the conidium which has been defined as "a specialised, non-motile, asexual propagule, usually caducous, not developing by cytoplasmic cleavage or free-cell formations" (see Kendrick, 1971). Conidia arise from conidiophores which may be single or highly differentiated structures, which function to position the developing conidium away from the medium and parent mycelium. Conidia may be formed in many ways and derive their name from the specific mechanisms i.e. blastospores, radulaspores, porospores, arthrospores and phialospores. For a fuller understanding of conidiospore formation see Kendrick (1971) and Weber & Hess (1976). The characteristic spore of the Aspergilli (and other economically important fungi such as *Penicillium* and *Trichoderma*) is the phialoconidium.

When conidia are produced in basipetal sequence from a phialide they are termed phialospores. The phialide is a conidiogenic cell in which the first conidium initially develops as an apical extension of the cell and is liberated by rupture or dissolution of the upper wall of the parent cell. From then on a basipetal succession of enteroblastic conidia may be produced in which the conidium wall does not derive from the conidiogenic cell wall. The successive production of conidia from a fixed conidiogenous locus is characteristic of the phialide; whereas in some fungi the conidia produced by the phialide may be shed individually, in others such as *Aspergillus* and *Penicillium* successively produced conidia are held together by an outer wall common to them all and as such form persistent chains of spores.

THE INFLUENCE OF THE ENVIRONMENT ON SPORULATION

It has long been recognised that growth and sporulation in fungi are generally incompatible and that the transition to reproduction is usually preceded by some environmental condition that checks vegetative growth. Almost all environmental parameters have been examined and in most cases can be shown to influence sporulation.

It is perhaps better to consider that vegetative growth and sporulation are cellular processes competing for limiting metabolic intermediates rather than as mutually exclusive phenomena. If vegetative growth and sporulation do not normally occur at the same time and are separated by definite metabolic shifts, the point of change may be associated with limitation of vegetative growth due to nutrient exhaustion or to some other environmentally controlled limiting process.

Recent studies by Axelrod and co-workers have shown that before a decision of commitment to sporulate is achieved there must be a fixed amount of biomass or colony growth. Axelrod (1972) and Axelrod, Gealt & Pastushok (1973) have demonstrated with *A. nidulans* that this point is reached by colonies grown from single conidia on surface culture and also by mycelium in liquid culture in about 20 h at 36°. Up to this time no exposure to a normal conidiating stimulus will result in conidiation whereas after this point the mycelium becomes receptive to stimuli which in turn will lead to conidiation approximately 7 - 8 h later. Similar observations have been made for microcycle conidiation with *Aspergillus niger*. Mature colonies of filamentous fungi growing on agar surfaces invariably have a zone of vegetative mycelium bordering the central area of conidiation. Since this purely vegetative area is approximately the distance travelled by the hyphal tips in 6 - 7 h it would appear that there is no latent period before competence is achieved by hyphae in a mature colony.

Surface cultivation techniques have yielded much useful information on the influence of the environment on the sporulation process. Yet the characteristic mycelial mat formation typical of surface cultures precludes any meaningful study of the biochemistry associated with sporulation. By using submerged cultivation techniques a more homogeneous truly filamentous mycelium can be produced. Sporulation in submerged culture can be routinely achieved by manipulation of the medium components and other environmental parameters. The most general condition for inducing sporulation is the reduction or exhaustion of assimilable nitrogen while carbohydrate is still present in the medium (Turian & Bianchi, 1972; Smith & Anderson, 1973).

Most studies on sporulation in filamentous fungi have utilized variations of flask culture technique which although ensuring adequate contact between mycelium and medium creates intrinsic disadvantages, in particular, poor conditions of aeration, variation between samples and inoculum size and difficulties in regulating factors such as pH. There is now a compelling need for developmental mycologists to examine many aspects of mycelial differentiation by way of simple and sophisticated fermenters where it becomes possible to grow within one vessel quantities of relatively similar mycelium under more rigorous experimental conditions such as aeration, temperature, agitation, pH control, incremental feeding and replacement cultures. However, a major disadvantage of submerged formation is the heterogeneous nature of mycelial form and any biochemical analysis can only be a summation of many physiological states. Some synchronous control of sporulation has been achieved with *Aspergillus niger* (Anderson & Smith, 1971a); *Achlya* (Griffin & Breuker, 1969) and *Penicillium digitatum* (Zeidler & Margaluth, 1972) using nutrient changes or cultural manipulations. The replacement method for *A. niger* sporulation has permitted a detailed examination of the biochemical changes associated with each stage of development (Smith & Ng, 1972; Ng, Smith & Anderson, 1972; Lloyd, Anderson, Smith & Morris, 1972; Ng, Smith & McIntosh, 1973a) (Fig. 2). It is interesting to note that this environmentally induced control of conidiation resembles in many ways some of the conidiation mutants found by Martinelli & Clutterbuck (1969, Martinelli & Clutterbuck, 1971). In particular, the *bristle* mutant of *A. nidulans* which forms immature conidiophores of indefinite length is similar to stage 1 in the replacement method

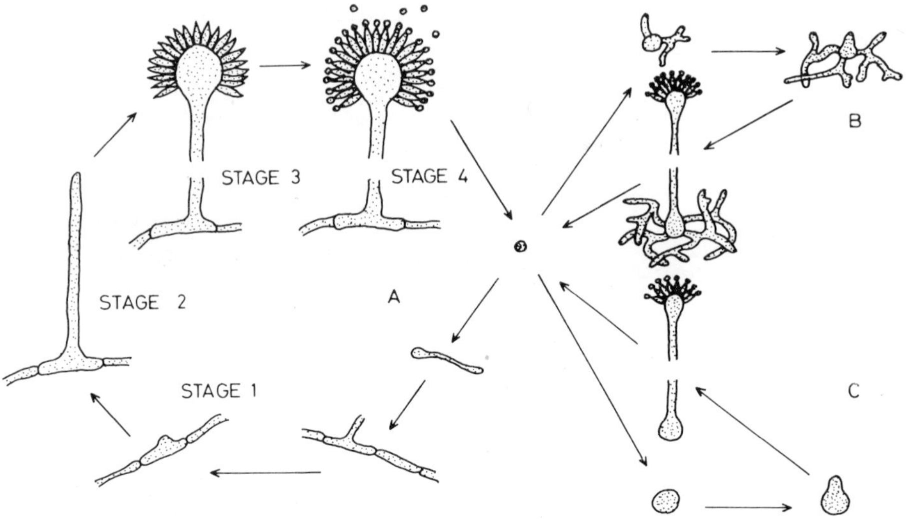

Fig. 2. Conidiation in Aspergillus niger (from Smith & Anderson, 1973).

of Anderson & Smith (1971a). Similarly, the abacus mutant is not unlike stage 2 where the vesicle and sterigma form but cannot proceed further unless they are replaced into stage 3 medium.

The medium environmental conditions within a batch system are in constant change and it is not possible to deduce from such a fermentation system whether sporulation is the result of nutrient limitation or from limitation of growth rate imposed by these conditions. Using chemostat culture it has been possible to show with A. niger that conidiation occurs by an interaction between growth rate and the ratio of the carbon and nitrogen sources in the medium (Ng, Smith & McIntosh, 1973b). The relationship between growth limitation and sporulation in filamentous fungi has only recently become understood and shows a remarkable similarity to many aspects of secondary metabolism (Bu'Lock, 1975). Ng et al. (1973b) have studied the effect of growth rate and the nature of several substrate limitations on conidiation in A. niger. Under citrate limitation conidiation intensity varied inversely with dilution rate. Conidiation was difficult to achieve under glucose limitation; poor conidiation occurred transiently at low dilution rates while at high rates the mycelium remained vegetative. It is probable that at high dilution rates catabolite repression prevails whereas at low dilution rates the substrate is supplied at too low a level and consequently autolysis will set in before conidiation can be established. Under ammonium limitation with citrate as the carbon source there was no conidiation. Nitrate limitation with citrate gave conidiophore initiation at low dilution rates but also early wash out. Good conidiation can be achieved in glucose limited nitrate medium.

Clearly these experiments varying dilution rate under citrate limitation demonstrate that conidiation can be controlled by growth rate while studies with glucose limitation further demonstrate that the composition of the medium is also important. Thus conidiation in _Aspergillus_ _niger_ is determined by the concentration and nature of the carbon and nitrogen sources in the culture medium. Limitation promotes conidiation - the type and character of this limitation is also critical.

Yet another way of studying sporulation in fungi is by means of the novel microcycle sporulation systems that have been developed in recent years. Microcycle sporulation can be defined as the immediate recapitulation of sporogenesis following spore germination and has been shown to occur in _Aspergillus_ _niger_ (Anderson & Smith, 1971a, b, 1972); _Neurospora_ _crassa_ (Cortat & Turian, 1974); _Penicillium_ _urticae_ (Sekiguchi _et_ _al_., 1975a, b, c) as well as some others (see Smith, 1977).

In most forms of microcycle sporulation there is an increase in the size of the spore by spherical growth, as a result of incubation at an elevated temperature. The role of temperature and spore development is more fully discussed by Anderson (1977) (see also Fig. 3). When the enlarged cells are exposed to a lower temperature they will produce a conidiating system. Microcycle conidiation has now been achieved in fermenter scale by Kuboye _et_ _al_. (1976) for _A_. _niger_. As a result it is now possible to examine this system in much greater detail. The giant cells of _A_. _niger_ are multinucleate and have exceptionally thick walls (Fig. 4). The advantage of the microcycle system over conventional myce-

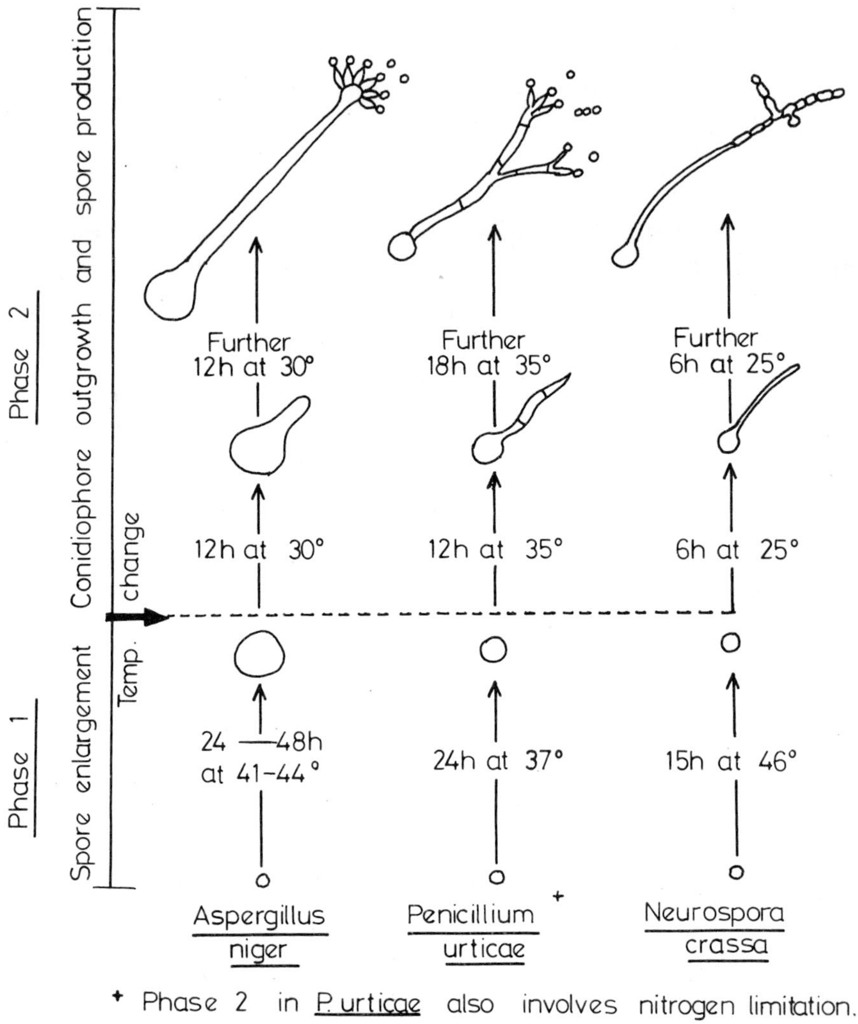

Fig. 3. The use of temperature to induce microcycle conidiation in filamentous fungi (from Anderson, 1977).

lial systems is primarily the absence of vegetative mycelium which may mask the changes occurring prior to and during sporulation.

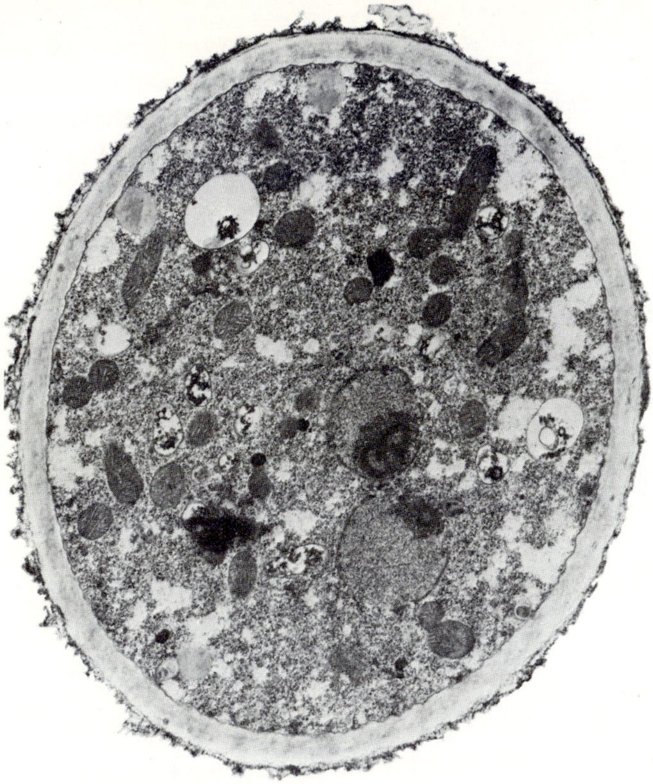

Fig. 4. The giant cell of Aspergillus niger.
Scale bar = 1 μm

ULTRASTRUCTURAL AND BIOCHEMICAL ASPECTS OF MICROCYCLE CONIDIATION IN ASPERGILLUS NIGER

Initial studies with the microcycle system in Aspergillus niger made use of test-tube culture. However, the development of a fermenter system has now permitted a deeper appreciation to be made of this system.

Large scale production of conidia for the initial fermenter inoculation is achieved by germinating conidia on cubes of sterile bread (Sansing & Ciegler, 1973). Harvesting the spores after several days' growth yields high numbers of a relatively asynchronous spore population. To increase the synchrony index a variety of methods were examined encompassing both induction synchrony and selection synchrony. However, washing the spores repeatedly in sterile distilled water was found to yield the greatest number of spores in a state that permitted a high degree of synchrony to be realised in the microcycle process.

The spores thus produced are introduced into a defined cultural medium (Anderson & Smith, 1971b, 1972) with the exception of the glucose level which has been increased to 15 g l^{-1}. Incubation of the cells at 44°C for 24 h yields the giant cell (average diameter 20 - 25 µm) and after this period if the temperature is dropped to 30°C conidiophore emergence is realised about 6 - 7 h later. Further incubation at 30°C permits complete conidiogenesis (Fig. 5). Membrane filtration enables the newly formed secondary conidia

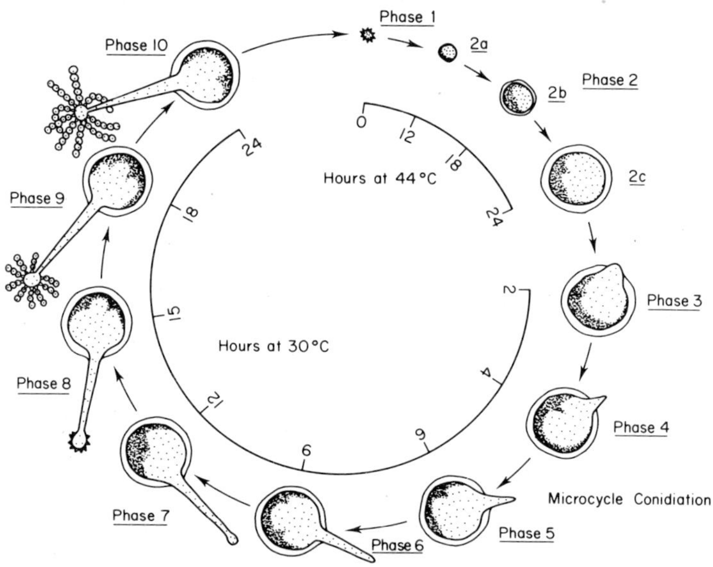

Fig. 5. Time clock of Aspergillus niger microcycle conidiation (Smith, 1977).

to be separated from the parental giant cell-conidiophore apparatus. Reinoculation of these secondary conidia into fresh medium at 44°C yields giant cells which on dropping the temperature to 30°C will emit conidiophores and conidia.

Thus this system makes an attractive alternative to conventional bulk spore production with the high number of spores produced (10^6 giant cells yield ~ 10^8 conidia ml^{-1}).

The fermenter system used was a Chemap 7 l laboratory scale instrument operated at a working volume of 5 l. Agitation was achieved by a double flat-bladed turbine operating throughout at 1500 rpm in the baffled reactor vessel. Two rotameters and a manually operated gas tap were used to control air and carbon dioxide flows to the fermenter. An input level of 5 % CO_2 was required to prevent germination of the giant cells at 44°C. The gas flow from the ring sparger below the impeller was adjusted by the rotameters to give 3.0 l min^{-1} air and 0.15 l min^{-1} CO_2. Foaming was controlled by the addition of 1.0 ml per 500 ml medium of silicone antifoam solution introduced as a 1 : 1 mixture with distilled water. The fermenter was inoculated with 2×10^6 conidia ml^{-1} and samples removed from the vessel by a built-in syphon system and examined microscopically.

The synchronous control of microcycle development in *Aspergillus niger* will undoubtedly permit a much more accurate appraisal of biochemical changes or fluctuations causally related to differentiation. In particular, the relationship between spherical growth and cylindrical growth is now being extensively studied together with nuclear behaviour within a multinucleate unicell, organelle biogenesis, nucleic acid regulation and also the production and utilisation of protoplasts from the giant cells.

Elevated temperature appears to be a specific feature of microcycle expression in *Aspergillus niger* as well as with most other examples (Anderson, 1977). The work of Cortat & Turian (1974) showed that exposure of *Neurospora crassa* to elevated temperature of 46°C for up to 15 hours followed by incubation at 25°C resulted in giant cell formation and eventual conidiogenesis. There are minor differences in the methods employed for achieving microcycle conidiation including a higher temperature range (42° - 47°C) than is possible with *A. niger*, shorter incubation period and higher spore density. Although both *N. crassa* and *A. niger* are capable of withstanding temperatures of 46°C and above, the subsequent development at lower temperatures is greatly inhibited. That an element of thermal denaturation of enzymic protein occurs at temperatures higher than 46°C is inevitable. Ojha & Turian (1968) illustrated with N. crassa that even at 40°C damage to the endoplasmic reticulum was in evidence under electron microscopy. Langvad (1972) noted that nuclear disruption, mitochondrial disintegration, granulation of cytoplasm, membrane fragmentation and ribosomal aggregation all resulted on exposure of Serpula lacrymans to a temperature 14°C above its optimal growth value (23°C). Both mitochondrial damage and nuclear disruption have also been reported in Rhizopus stolonifer and Monilia fructicola on incubating spores for a brief period at 52°C (Baker & Smith, 1970). Cortat & Turian (1974) consider the expression of microcycle in terms of premature conidiation since in their model system conidiation (proconidia in this scheme) commences around 6 h at the lower incubation temperature of 25°C. Addition of certain amino acids notably glutamic and serine to the sporulation medium with Penicillium digitatum (Zeidler & Margaluth, 1972, 1973) enhanced both sporulation and microcycle conidiation while tryptophan caused conidial chain elongation. The incubation temperature was 26°C. Sekiguchi et al. (1975a, b, c) achieved microcycle conidiation in Penicillium urticae by incubation at 37°C for 24 h followed by up to 30 h at 35°C. This scheme, however, employs a medium-replacement step at the point of temperature lowering. The elevated temperature stage forming giant cells (6.0 - 6.5 μm) from normal untreated glucose medium. Filtration and resuspension in a cysteine-glucose medium, nitrogen poor, is required for microcycle conidiation expression. This method of cultivation enables 50 % of the original spore inoculum to produce conidia directly from an emergent conidiophore. For a fuller discussion of temperature and growth see Anderson (1977).

WALL SYNTHESIS IN MICROCYCLE CONIDIATION OF A. niger

The shape and form of fungi is determined largely by the presence of the rigid outer cell wall. The fungal cell wall is a complex and structured assembly of polymers in which the chemical nature of the polymers and the spatial relationship of the polymers to each other can alter during growth and differentiation (Bartnicki-Garcia, 1968, 1973; Rosenberger, 1976). There have been only a few analytical studies of the chemical and physical changes of wall structure related to sporulation in filamentous fungi (for references see Smith, 1977).

The extreme thickness of the giant cell wall of Aspergillus niger (Smith et al., 1976) has prompted a special study of this aspect together with wall changes during conidiophore emission (Deans & Smith, 1977a). Following the fractionation scheme of Mahadevan & Tatum (1965) the main cell wall components change as follows during conidiophore formation. The two glucan fractions plus chitin constitute the major polysaccharide components of the cell wall, the alpha-glucan fraction containing 73 % alpha 1-3 glucan. The initial period of spherical growth, up to around 12 hours, sees a build up in alpha-glucans in the cell wall. However, between 12 and 16 hours, a noticeable drop in alpha-glucan content is observed, but this does not persist, and the level again increases to a maximum just prior to the elevated incubation temperature being lowered to 30°C. The values obtained during the microcycle (30°C) phase illustrate an apparent rapid utilisation of alpha-glucans. Since only alpha-glucans behave in this manner, all other carbohydrate wall fractions continuing to increase during conidiophore emergence and conidiogenesis, it is believed the alpha-glucans act as storage compounds during the earlier phases of growth. Beta-glucans appear to be an equally important contributory group, being of approximately the same percentage as alpha-glucans. However, there is no indication of these beta-glucans being deployed as storage compounds, their function being more of a structural nature.

Chitin levels increase steadily throughout the cycle. The peaks of activity of the synthetic enzyme, chitin synthase, occur around 11 - 13 hours at 44°C, and around 12 hours at 30°C (unpublished results). Chitin is known to be a very important structural element in the cell wall, forming rodlets embedded in the glucan matrix. This role is supported by the high levels observed as the cycle passes through spherical growth into conidiophore emergence and conidiation, both processes requiring considerable cell wall synthesis. That the beta-glucan levels closely parallel those of chitin, whereas alpha-glucan levels decrease, would suggest that alpha-glucans are being utilised to a greater extent in the formation of the cell wall amorphous matrix. The heterogeneous glucose-galactose-mannan fraction does not appear to contribute greatly to the wall fabric, the levels seldom rising above 15 %, but the steady overall increase during the 48 hour cycle suggests incorporation within the wall matrix, there being only a small recorded drop in level of incorporation during hours 10 - 15 at 44°C.

The levels of both protein and lipid have interesting profiles. The protein increase during the first 16 hours at 44°C is to be expected, since during this time most of the enzyme quota of the cell is being amassed, spherical growth is well established, and maximum wall formation is occurring. After this time, there is an equally rapid fall-off in protein content within the wall, spherical growth beginning also to decrease in rate of expansion. The renewed upward trend at 30°C is due mainly to conidiophore emergence, wall synthesis again being one of the most important cellular functions, commanding a great deal of the cell's resources. Conidiogenesis requires a large amount of synthetic activity, each conidium requiring to be endowed with its own wall forming apparatus. Hence, the level of cell wall protein increases, but it is noted, not at as fast a rate as the original germination period of spherical growth.

Cell wall lipid levels fall dramatically during the first 16 hours incubation at 44°C. Utilization by the rapidly growing cell seems a logical explanation for this drop in lipid, maximum activity within the cell, as stated above, appears within this same time. The cell, having achieved close to its maximal diameter, complete enzyme complement, and an alternative storage compound to be degraded in the alpha-glucans, can again turn to synthesising and storing wall lipid. This synthetic activity occurs up until the temperature is dropped to 30°C, when again reutilisation of these reserves takes place. For 16 hours at 30°C, lipid is utilised as an energy source, but with incipient conidiogenesis, fat storage is observed, the levels representing the conidial values.

ORGANELLE BIOGENESIS

As a prelude to an extensive examination of the ener-

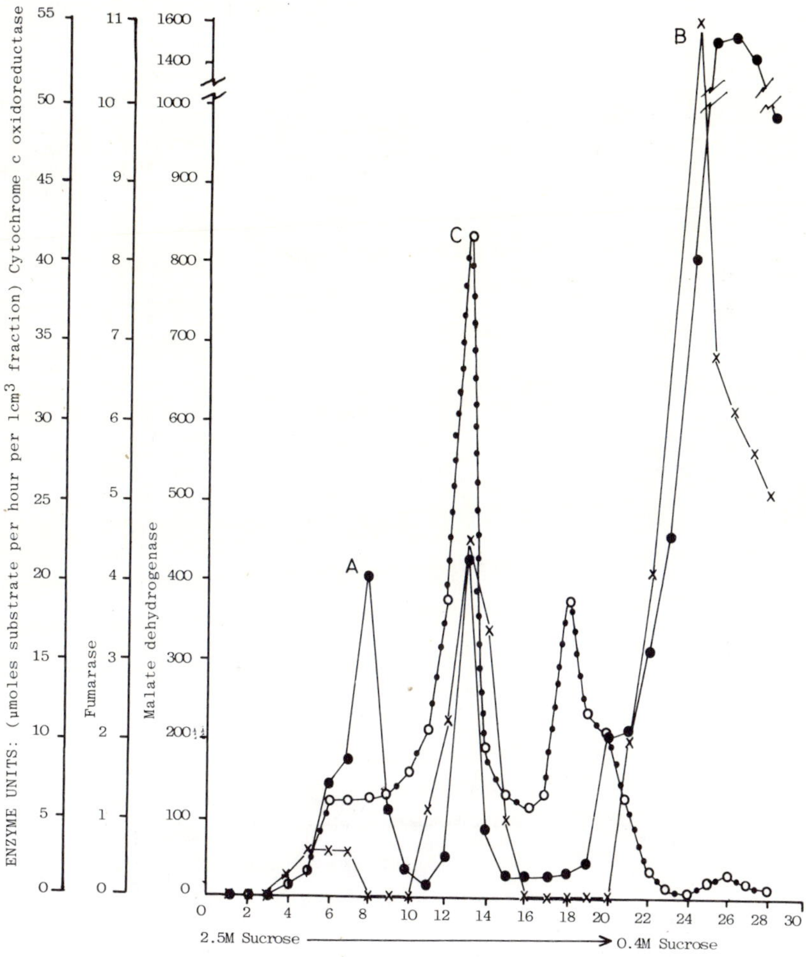

Fig. 6. Distribution of marker enzymes after linear sucrose density gradient centrifugation of cell extracts of *Aspergillus niger* grown under microcycle conidiation conditions. 1 ml fractions were collected from the bottom of the centrifuge tube and were numbered in order of collection. A, malate dehydrogenase (●-●), B, fumarase (x-x) and C, cytochrome *c* oxidoreductase (o-o) (from Davis & Smith, 1977).

getics of microcycle conidiation and to the role of mitochondria in this process several methods were examined with a view to obtaining intact organelles from the giant cell. Conventional methods for cell disruption for organelle extraction were not highly successful. However, a cell smash method

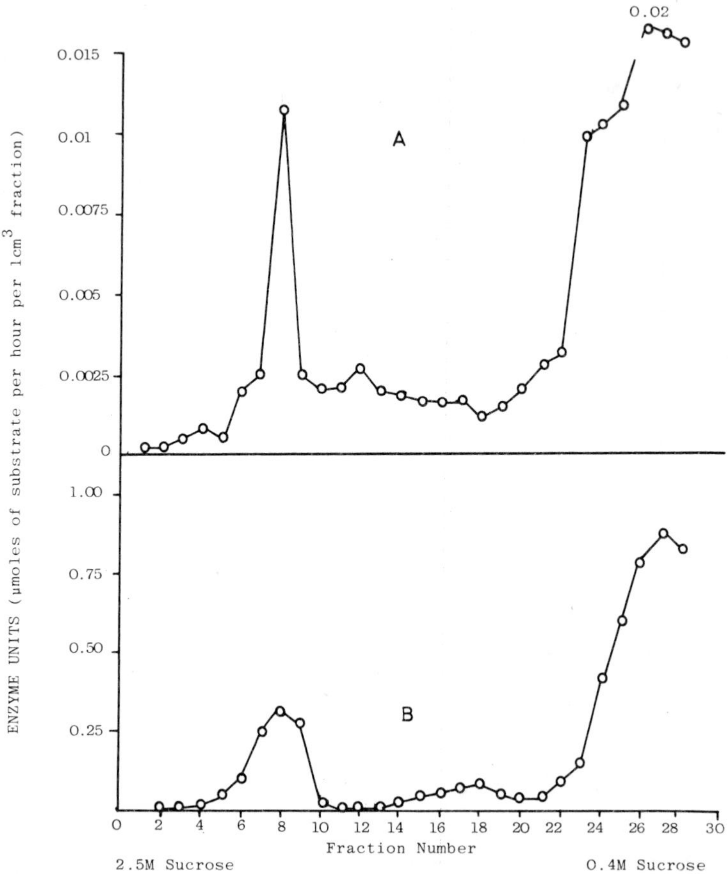

Fig. 7. Distribution of glyoxylate cycle enzymes after linear sucrose density gradient centrifugation of cell extracts of *Aspergillus niger* grown under microcycle conidiation conditions. 1 ml fractions were collected from the bottom of the centrifuge tube and were numbered in order of collection. A, isocitrate lyase and B, malate synthase (from Davis & Smith, 1977).

utilising the Braun MSK rotary homogeniser has been successfully developed (Davis & Smith, 1977) and intact mitochondria have been isolated (Fig. 6). The organelles were characterized using organelle specific marker enzymes. Fumarase was used as a mitochondrial localisation of the enzyme. Glyoxosomal enzymes were measured but their localisation in a microbody or glyoxosome were inconclusive (Fig. 7). Unpublished studies have shown that when giant cells of A. niger undergo starvation there is considerable ultrastructural evidence of microbody formation.

PROTOPLAST FORMATION

Osmotically sensitive protoplasts have been liberated enzymatically from giant cells of A. niger by utilising an induced extracellular enzyme system isolated from culture filtrates of Trichoderma viride grown on a medium supplemented with purified wall polysaccharides of A. niger giant cells. Protoplast release occurred with complete digestion of the cell wall. A complete characterisation of these protoplasts has been carried out (Davis et al., 1977). At present it has not been possible to get regeneration of these protoplasts which may indicate some damage to the protoplast at the high incubation temperature. Furthermore, in all giant cell studies the digestive enzyme must be supplemented with commercial chitinase to achieve dissolution of the cell wall. The synchronous nature of this system will allow a systematic analysis of organelle biogenesis throughout conidiation.

THE INFLUENCE OF METABOLIC INHIBITORS ON MICROCYCLE CONIDIATION

A range of metabolic inhibitors was employed to determine the relationship between cellular inhibition of particular metabolic functions and expression of this inhibition in morphogenesis. The following compounds were used, the abbreviations accompanying each indicating the generally accepted mode of metabolic action (UOP, uncoupler of oxidative phosphorylation; PS, inhibitor of protein synthesis; MF, inhibitor of mitochondrial function; NAB, inhibitor of nucleic acid biosynthesis) (Burke & Germond, 1976; Boehringer Program of Inhibitors, 1976): 2, 4 : dinitrophenol (UOP); puromycin (PS); tetracycline (MF and PS); rifampicin (MF and NAB); cycloheximide (PS); sodium azide (UOP); chloramphenicol (MF); emetine (PS); 5-fluorouracil (NAB); actinomycin D (NAB); 8-azaguanine (NAB); and mitomycin C (NAB). The inhibitors were added, at time of inoculation, to test tubes containing 9 ml medium (Anderson & Smith, 1971, 1972) with a 2.0×10^6 ml^{-1} conidia, and incubated at 44°C for 24 h. Cell diameter was measured at the end of this period. Dropping the temperature to 30°C allows microcycle conidiation to proceed, and the percentage of the population actively sporulating after 24 h at this temperature was monitored. These experiments gave dosage curves of inhibition (Fig. 8). Typically, it was found that conidiation was more susceptible to inhibition than giant cell diameter. This is shown at concentrations where the cell diameter remains 23 μm, yet conidiation is either greatly reduced or totally absent (compare mitomycin c 120 μg ml^{-1} and actinomycin D 30 μg ml^{-1}). Puromycin (protein synthesis inhibitor, Pestka, 1971) elicits a response where the diameter is apparently unaffected up to 60 μg ml^{-1}, yet conidiation ceases at concentrations of 40 μg ml^{-1} and higher. One feature of puromycin inhibition was the production of extra thick cell walls. Thus, the prolonged large diameter is no longer due to overall cytoplasmic enlargement, but merely wall thickening. This may represent the lytic element of wall growth being depressed, permitting synthetic events to go unchecked. Rifampicin also illustrates a feature of interest wherein cells of 20 μm are able to support healthy conidiophore/conidia production, yet when the diameter is reduced slightly more (100 μg ml^{-1} level) total inhibition of conidiation is observed.

In a second experiment, 4 inhibitors were chosen for closer study, at specific concentrations. Actinomycin D, chloramphenicol, cycloheximide and tetracycline were added, individually, at 2 hourly intervals and either remained in contact with the cells for the remainder of the 44°C incubation regime, or were removed by centrifugation after 2 hours contact. This enabled a more accurate pinpointing of the time at which the inhibitors elicited their effect. Additions of

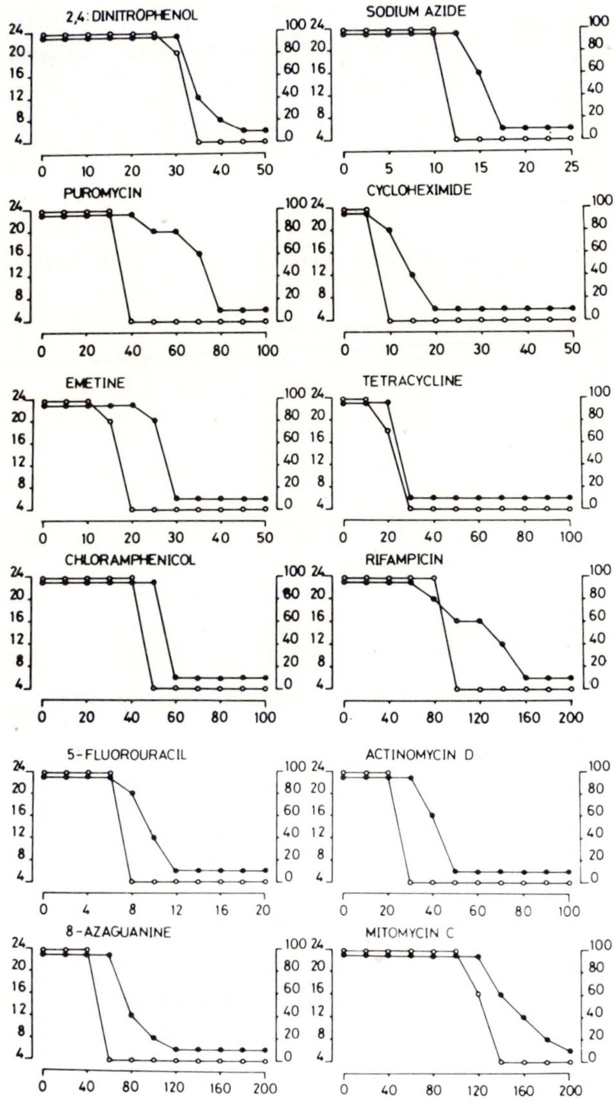

Fig. 8. The influence of metabolic inhibitors on microcycle conidiation of *Aspergillus niger* (Deans & Smith, 1977b).
Abscissae: inhibitor concentration, g l^{-1};
Ordinates: giant cell diameter, μm (●-●);
conidiation, % (o-o).

inhibitor with no subsequent removal causes a greater inhibitory effect than in the situation where the inhibitor is only in cell contact for 2 h. Actinomycin D addition, without removal, is inhibitory upon giant cell diameter up until 14 - 16 h at 44°C; chloramphenicol up until 8 - 10 h at 44°C; cycloheximide up until 14 h at 44°C; and tetracycline up until 14 - 16 h at 44°C. Removal of the inhibitor after 2 h contact shows a different inhibitory pattern. Here with actinomycin D, inhibition is in evidence only up until addition after 10 h at 44°C. Chloramphenicol and tetracycline are similar in that up to until 10 h at 44°C, addition and removal causes inhibition, whereas addition thereafter has virtually no effect upon cell diameter. Cycloheximide, however, retains its potent inhibitory effect up to 14 - 16 h at 44°C. The inhibitory effect is lessened by removal after2 h contact. In a few cases, where there is no great distinction between removal/non removal inhibitory patterns (seen in later samples $>$ 18 h at 44°C) inhibitor retention may be occurring. When the cell is large, $>$ 18 μm, the thick cell wall may be playing some role in keeping the inhibitor within the confines of the cell, permitting the inhibitory effect to be prolonged. Subsequent development, microcycle conidiation, is related to the diameter achieved during giant cell growth phase at 44°C. The point at which an inhibitor ceases to exert an effect upon diameter does not imply that no further inhibition of conidiation will occur, since in some cases, as with dose response curves, a giant cell of $>$ 21 μm is incapable of supporting conidiogenesis. Again, these cells exposed to inhibitor, with no removal, are more inhibited in their subsequent development.

The decoding of DNA to RNA, inhibited by actinomycin D, seems completed prior to 8 h at 44°C. The compounds responsible for inhibiting mitochondrial function, notably chloramphenicol and tetracycline, are active against oxidative phosphorylation and prevent, by energy starvation, developmental progression. These inhibitors when added to the culture prior to 10 h at 44°C will restrict conidiophore production, and therefore suggest that processes of conidiophore outgrowth, elongation and conidiogenesis are all demanding in energy, and cannot proceed with impaired mitochondrial function.

REFERENCES

Anderson, J. G. (1977). Temperature induction of fungal development. In: The filamentous fungi, Vol. 3 in press. Edited by Smith, J. E. & Berry, D. R. London, Edward Arnold Publisher.

Anderson, J. G. & Smith, J. E. (1971a). Synchronous initiation and maturation of *Aspergillus niger* conidiophores. Transactions of the British Mycological Society 56, 9-29.

Anderson, J. G. & Smith, J. E. (1971b). The production of conidiophores and conidia by newly germinated conidia of *Aspergillus niger* (microcycle conidiation). J. Gen. Microbiol. 69, 185-197.

Anderson, J.G. & Smith, J. E. (1972). The effects of elevated temperatures on spore swelling and germination in *Aspergillus niger*. Can. J. Microbiol. 18, 289-297.

Axelrod, D. E. (1972). Kinetics of differentiation of conidiophores and conidia by colonies of *Aspergillus nidulans*. J. Gen. Microbiol. 73, 181-184.

Axelrod, D. E., Gealt, M. & Pastushok, M. (1973). Gene control of developmental competence in *Aspergillus nidulans*. Developmental Biology 34, 9-15.

Baker, W. L. & Smith, J. E. (1970). Heat induced ultrastructural changes in germinating spores of *Rhizopus stolonifer* and *Monilia fructicola*. Phytopathology 60, 869-874.

Bartnicki-Garcia, S. (1968). Cell wall chemistry, morphogenesis and taxonomy in fungi. Annual Review of Microbiology 22, 87-108.

Bartnicki-Garcia, S. (1973). Fundamental aspects of hyphal morphogenesis. Symposium Soc. Gen. Microbiol. 23, 245-267.

Boehringer Program of Inhibitors (1976).

Bu'Lock, J. D. (1975). Secondary metabolism in fungi and its relationship to growth and development. In: The filamentous fungi, Vol. 1, pp. 33-58. Edited by Smith, J. E. & Berry, D. R. London, Edward Arnold Publishers.

Burke, D. O. & Germond, M. (1976). Effect of inhibitors on the morphology and growth of Allomyces macrogynus. J. Gen. Microbiol. 95, 213-219.

Carmichael, J. W. (1971). Blastospores, aleurospores, chlamydospores. In: Taxonomy of fungi imperfecti, pp. 50-70. Edited by Kendrick, B. Toronto and Buffalo, University of Toronto Press.

Clutterbuck, A. J. (1969). A mutational analysis of conidial development in Aspergillus nidulans. Genetics 63, 317-327.

Clutterbuck, A. J. (1977). The genetics of conidiation in Aspergillus nidulans. In: The physiology and genetics of Aspergillus. In Press. Edited by Smith, J. E. & Pateman, J. London, Academic Press.

Cortat, M. & Turian, G. (1974). Conidiation of Neurospora crassa in submerged culture without mycelial phase. Arch. Microbiol. 95, 305.

Davis, B. & Smith, J. E. (1977). Organelle biogenesis in Aspergillus niger. FEMS Letters, 1, 51-53.

Davis, B., d'Avillez-Paixao, M. T., Deans, S. G. & Smith, J. E. (1977). Protoplast formation from giant cells of Aspergillus niger. Transactions British Mycol. Soc. in press.

Deans, S. G. & Smith, J. E. (1977a). Wall biogenesis in microcycle conidiation of Aspergillus niger. Manuscript in preparation.

Deans, S. G. & Smith, J. E. (1977b). The influence of metabolic inhibitors on microcycle conidiation in Aspergillus niger. Transactions British Mycol. Soc. submitted.

Griffin, D. H. & Breuker, C. (1969). Ribonucleic acid synthesis during the differentiation of sporangia in the water mold Achlya. J. Bacteriol. 98, 689-696.

Kendrick, B. (1971). Ed. Taxonomy of fungi imperfecti. Toronto and Buffalo, University of Toronto Press.

Kuboye, A. O., Anderson, J. G. & Smith, J. E. (1976). Control of autolysis of a spherical cell form of

Aspergillus niger. Transactions British Mycol. Soc. 67, 27-31.

Langvad, F. (1972). The effect of supraoptimal temperature on the fine structure of *Mesulius lacrymans*. J. Gen. Microbiol. 70, 157-159.

Lloyd, S., Anderson, J. G., Smith, J. E. & Morris, E. O. (1972). The effect of medium nitrogen and of conidiation on esterase synthesis in *Aspergillus niger*. Transactions British Mycol. Soc. 59, 63-70.

Lovett, J. S. (1975). Growth and differentiation in the water mold *Blastocladiella emersonii*: cytodifferentiation and the role of ribonucleic acid and protein synthesis. Bacteriological Reviews 39, 345-404.

Mahadevan, P. R. & Tatum, E. L. (1965). Relationship of the major constituents of the *Neurospora crassa* cell wall to the wild type and colonial morphology. J. Bacteriol. 80, 1073-1081.

Martinelli, S. D. & Clutterbuck, A. J. (1971). A quantitative survey of conidiation mutants in *Aspergillus nidulans*. J. Gen. Microbiol. 69, 261-268.

Ng, A. M. L., Smith, J. E. & McIntosh, A. F. (1973a). Changes in activity of tricarboxylic acid cycle and glyoxylate cycle enzymes during synchronous development of *Aspergillus niger*. Transactions British Mycol. Soc. 61, 13-20.

Ng, A. M. L., Smith, J. E. & McIntosh, A. F. (1973b). Conidiation of *Aspergillus niger* in continuous culture. Arch. Microbiol. 88, 119-126.

Ng, W. S., Smith, J. E. & Anderson, J. G. (1972). Changes in carbon catabolic pathways during synchronous development of *Aspergillus niger*. J. Gen. Microbiol. 71, 495-504.

Ojha, M. N. & Turian, S. (1968). Thermostimulation of conidiation and succinic oxidative metabolism of *Neurospora crassa*. Arch. Microbiol. 63, 232-243.

Pestka, S. (1971). Inhibitors of ribosomal functions. Ann. Rev. Biochem. 40, 697-710.

Rosenberger, R. F. (1976). The cell wall. In: The filamentous fungi, Vol. 2, pp. 328-344. Edited by Smith, J. E. & Berry, D. R. London, Edward Arnold Publishers.

Sansing, G. A. & Ciegler, A. (1973). Mass propagation of conidia from several Aspergillus and Penicillium species. Appl. Microbiol. 26, 830-831.

Sekiguchi, J., Gaucher, G. M. & Costerton, J. W. (1975a). Microcycle conidiation in Penicillium urticae. An ultrastructural investigation of spherical spore growth. Can. J. Microbiol. 21, 2048-2058.

Sekiguchi, J., Gaucher, G. M. & Costerton, J. W. (1975b). Microcycle conidiation in Penicillium urticae. An ultrastructural investigation of conidial germination and outgrowth. Can. J. Microbiol. 21, 2059-2068.

Sekiguchi, J., Gaucher, G. M. & Costerton, J. W. (1975c). Microcycle conidiation in Penicillium urticae. An ultrastructural investigation of conidiogenesis. Can. J. Microbiol. 21, 2069-2083.

Smith, J. E. (1977). Asexual sporulation in filamentous fungi. In: The filamentous fungi, Vol. 3, in press. Edited by Smith, J. E. & Berry, D. R. London, Edward Arnold Publishers.

Smith, J. E. & Anderson, J. S. (1973). Differentiation in the Aspergilli. Symposium Soc. Gen. Microbiol. 23, 295-337.

Smith, J. E. & Berry, D. R. (1974). An introduction to biochemistry of fungal development. London, Academic Press.

Smith, J. E. & Galbraith, J. C. (1971). Biochemical and physiological aspects of differentiation in the fungi. Advances in Microbial Physiology 5, 45-134.

Smith, J. E. & Ng, W. S. (1972). Fluorometric determination of glycolytic intermediates and adenylates during sequential change in replacement culture of Aspergillus niger. Can. J. Microbiol. 18, 1657-1664.

Smith, J. E., Gull, K., Anderson, J. G. & Deans, S. G. (1976). Organelle changes during fungal spore germination. In: The fungal spore: form and function. pp. 301-384. Edited by Weber, D. J. & Hess, W. M. New York, John Wiley & Sons.

Turian, G. (1974). Sporogenesis in fungi. Ann. Rev. of Phytopathology 12, 129-137.

Turian, G. (1975). Differentiation in Allomyces and Neuro-

spora. Transactions British Mycol. Soc. 64, 367-380.

Turian, G. (1976). Spores in Ascomycetes, their controlled differentiation. In: The fungal spore: form and function. pp. 715-788. Edited by Weber, D. J. & Hess, W. M. New York, John Wiley & Sons.

Turian, G. & Bianchi, D. E. (1972). Conidiation in *Neurospora*. Botanical Reviews 38, 119-154.

Weber, D. J. & Hess, W. M. (1976). Eds. The fungal spore: form and function. New York, John Wiley & Sons.

Zeidler, G. & Margaluth, P. (1972). Synchronized sporulation in *Penicillium digitatum* (Sece.). Can. J. Microbiol. 18, 1685-1690.

Zeidler, G. & Margaluth, P. (1973). Modification of the sporulation cycle in *Penicillium digitatum*. Can. J. Microbiol. 19, 481-483.

FORMATION OF BASIDIOCARPS

G. Manachère

During their life cycle, many superior fungi, such as Ascomycetes and Basidiomycetes present structures producing their sexual spores, generally qualified as "carpophores", "fruit-body" or "Fruchtkörper". Here, we will only be concerned with problems relating to the formation of Basidiomycetes carpophores. Nevertheless, it may be kept in mind that the morphogenesis of Ascomycetes carpophores, or ascocarps, shows analogies to basidiocarp development.

Later on, the knowledge that we have of basidiocarp formation will be considered firstly from the ontogenic and then the physiological point of view.

BASIDIOCARP FORMATION FROM THE ONTOGENIC POINT OF VIEW

For a number of years, intensive research has been carried out to define morphological, histological and cytological characteristics of various types of basidiocarps. At this stage, it would be useful to cite the exemplary works of Corner (cf. Burnett, 1968; Reijnders, 1963) concerning Agaricales, Polyporales and Clavariaceae as well as the fundamental works of Reijnders (1963) and Kühner (1926) essentially devoted to Agaricales.

Numerous carpophores (basidiocarps as well as ascocarps) produce, at maturity, a characteristic structure bearing sexual spores or hymenium, the latter constituted by specific cells (basidium or ascus) arranged in a regular and parallel manner, among which are eventually intermingled other non-sporogenous cells (e.g. cystidia). The hymenium, composed of apical cells, is either covered by a sterile tissue or exposed to fresh air by a tissue also sterile. It is the form of these sterile tissues (particularly variable according to systematic groups) which determines the morphology of the adult carpophore.

Searchers have, for a long time, tried to reply to the following question: "does a basidiocarp arise from one hyphal cell of a potentially fruitful mycelium or does it, on the contrary, result from the aggregation of branches issued from various adjoining cells?"

Brefeld (1877) considered that the primordia of Coprinus stercorarius could be originated from only one cell. Buller (1933) confirmed this point of view for the same species. Recently HuguenEy (in Manachère et al., 1974) affirmed a similar origin for Coprinus congregatus; however, after meticulous "in vivo" observations, Matthews & Niederpruem (1972) concluded that the primordia of Coprinus lagopus did not originate from branches of an unique cell but from "activation centres", resulting from the interaction of cells and produced by the secondary and tertiary branching systems of the hyphae of original mycelium. It is therefore evident that the problem of the primordium origin, unicellular or not, does not actually have a solution for a great number of species.

Whether their origin is uni- or pluricellular, primordial hyphae gather to constitute the carpophore bud. Reijnders qualifies the tissue of such a primordium as protenchyme, if no other differentiation has occurred. However, Reijnders also underlines the practical difficulty in recognizing a protenchyme, sensu stricto, which would effectively be constituted of non-differentiated hyphae. The following data relative to the development of the Agaricales carpophores, Coprinus congregatus, studied in our laboratory, permits the understanding of difficulties already evoked. When they are cultivated in darkness, at a temperature of about 25°C, the vegetative mycelia of this mushroom do not produce basidiocarps, not even buds. However, when mycelia are illuminated 12 hours per day after a preliminary stay in darkness, they generally bear mature carpophores at a "0-hour stage", after five or six days, more rarely after four or seven days. This stage can be observed at the beginning of photoperiods. It is possible to recognize and define by their morphology and physiological characteristics the stage of development preceding 0-hour stage, noticeable for example at the beginning and end of photoperiods and qualified as minus 24-hours stage, minus 36-hours stage, minus 60-hours stage ... etc ... (Manachère, 1967; 1970). As already indicated, just after the preliminary stay in darkness, the C. congregatus mycelia do not carry any recognizable primordia. However, in more favourable cases, young buds are already visible approximately 10 hours after the start of the first photoperiod. In the meaning that we use they can correspond to minus 110-hours stage (Huguenev, in Manachère et. al., 1974). By branching, young buds rapidly take a characteristic bush aspect. The terminal articles of certain hyphae of these primordia present the aspect and physiological comportment of "hairs" normally characterizing the pilei of that type of carpophore. In other words, these "differentiated" cells can be seen very early and one could consider that the hyphae of primordium, at the minus 110-hours stage, do not properly constitute the protenchyme. In the interior of primordia one can see, at the minus 84-hours stage, binucleate apical articles, fence-ordered, and forming the outline of future hymenium mainly composed of basids, which are destinated to produce later four haploid basidiospores or spores of sexual reproduction. From our point of view, it is considered that it is not until this hymenial outline is recognizable that the primordium has really left its protenchymatic stage. From the morphological aspect it is only after the minus 72-hours stage that the pilei caps become perceptible to the naked eye. The primordium is therefore globular and its diameter is included between one-half and one mm. At the minus 36-hours stage the buds are of a size varying from 1.5 mm to 5.5 mm; the basids are still binucleate. At the minus 24-hours stage the carpophores reach a size between 4.5 mm and 10 mm; moreover in most of the basids, a characteristic nuclear fusion occurs constituting the first stage of meiosis. At the minus 12-hours stage the edges of pilei are free from the stipe; the majority of basids contain four nuclei each of n chromosomes. During the hours that follow, these nuclei migrate into the young basidiospores produced at apical tips or sterigmata developed at the free surface of basids. Between the minus 12-hours stage and 0-hour stage primordia are characterized by a spectacular elongation. From a size of 10 - 20 mm they reach a height of 40 - 80 mm during these corresponding 12 hours. At the zero-hour stage the carpophores present open pilei, darkening and then projecting their basidiospores and finally undergo a rapid autolysis which progresses from their free edges and stops after 3 - 4 hours.

As described by authors such as Kühner & Reijnders, the ontogenesis of C. congregatus carpophores constitutes the development of an angiocarpic type. According to the definition proposed by Reijnders, an angiocarpic primordium is a young carpophore "where the nascent hymenophore, or the part where the hymenophore will later be differentiated, is covered by a tissue during a certain period of the development". Reijnders distinguished primary angiocarpy from secondary angiocarpy; the first "is characterized because the pileus edge, the hymenophore and sometimes the pileus and the stipe become differentiated above the surface of the protenchyme"; as in C. congregatus and various other Coprinus, the development of the pileus is superficial, even if it is endogenous, when for other species as in Agaricus, such as the cultivated mushroom A. bisporus, it is relatively deep. Secondary angiocarpy appears, said Reijnders, "when the hyphae from a surface already differentiated (from the pileus or the stipe) grow lopsided to the exterior and cover the primordium or a part of it". This is the case for the Lentinus genus and certain Boletus.

Schematically, angiocarpy is the opposite to another form of development, qualified as gymnocarpy. According to Reijnders, a gymnocarpic primordium is "a primordium where, from the beginning, the hymenophore is naked and therefore the place where the hymenophore was differentiated had never been covered by a tissue". To this definition corresponds the carpophore type of Cantharellus and Pleurotus species. It is evident that numerous cases of secondary angiocarpy are derived from a situation primitively gymnocarpic.

It is indispensable to insist on the fact that the ontogenic schemas proposed above are essentially obtained from fragmentary observations of the developmental stages of carpophores collected in nature, more often accidentally. It would be preferable that a searcher cultivates a mycelium able to produce carpophores in a controlable manner, which could be set aside and observed at successive periods of development. Here, new genetic and physiological problems arise: the controlled culture of mycelia, with the object of basidiocarp production, necessitates the control of endogenous and exogenous factors, firstly responsible for the production of these structures and then for their development until maturity.

BASIDIOCARP FORMATION FROM THE PHYSIOLOGICAL POINT OF VIEW

Two phases can be distinguished in basidiocarp development. An initial phase is the release of fructification from a vegetative mycelium. This phase is often qualified as the "initiation phase" or "carpogenesis" see Reijnders (1963); the first phase is followed, without any recognizable gap, by the morphogenetic phase during which are elaborated various structures, more or less differentiated, of the basidiocarps.

1. The "Initiation" Phase

The realisation of this phase is simultaneously dependent on "endogenous" factors and various environmental or "exogenous" factors. We will envisage, successively, examples of intervention of these two kinds of factors.

Role of "Endogenous" Factors. The initial phase and consequently the ulterior morphogenesis of basidiocarps, can only be accomplished if certain genetic conditions are fulfilled (Esser, 1977).

Recent studies, of which full generalisation is not yet possible, envisage the participation of cyclic AMP in the initiation of fructification. Thus, Uno & Ishikawa (1971, 1973, 1974) established that cell extracts of carpophores or dicaryotic mycelia of different species, could induce the production of basidiocarps from monocaryotic mycelia of *Coprinus macrorhizus*, which normally do not produce primordia. These authors attributed the induction power of the extracts used to a "fruiting-inducing substance" or "F.I.S.". They identified this F.I.S. with cyclic AMP and with a protein to which it is bound. In fact the monocaryotic mycelium of a mutant "fis^{C}" of C. macrorhizus produces carpophores in light but does not form them in darkness. Parallely, such a mycelium produces cyclic AMP when illuminated but not in darkness. In the case of Coprinus lagopus, Matthews & Niederpruem (1972) established that the addition of cyclic AMP to the culture medium, minimal in glucose, advanced the primordia production by 24 hours.

The findings given above permit us to understand that, if these "endogenous" factors (genetic factors, substances produced by fungi like F.I.S. and others cf. Eger, 1965a, b; 1968) have a capital importance in the release of fructification, then various "exogenous" factors also have a primordial role.

Role of "Exogenous" Factors.

a) Nutritional Factors. It is only with the utmost prudence that it is possible to actually interprete certain isolated results and, a fortiori, to generalize them (cf. Volz & Beneke, 1969). Thus, different authors have indicated that an excess of glucose inhibits the fructification of certain species. It is however necessary to mention that glucose can, eventually, exert an inhibitory effect only if certain other constituents of the nutritional medium intervene as limiting factors. Thus, in the case of C. congregatus, the inhibition due to glucose can be suppressed by a suitable complementary supply of asparagine. But the nature of carbon or nitrogen sources, incorporated into the medium, can be involved in the fructification initiation, reactions varying from one species to another (Bille-Hansen, 1953b; Hawker, 1966). The presence of asparagine could prevent the reproduction of diverse Basidiomycetes (C. sassii, Bille-Hansen, 1953a; Sphaerobolus stellatus, Alasoadura, 1963) but favour that of others (Merulius lacrymans, Zur Lippe & Nesemann, 1959; C. congregatus - under certain conditions - Manachère, 1970). In addition, the role of carbon/nitrogen ratios must be taken into consideration; therefore, in C. congregatus no visible bud is produced by cultivated mycelia on nutritional media, where this ratio is superior to 100, the principal source of carbon being glucose and that of nitrogen, asparagine (Horrière, 1976). As always, from the nutritional point of view, we must underline the importance of thiamine or vitamin B1, necessary for the induction of carpogenesis (Hawker, 1942). According to Wessels (1965), this vitamin could be responsible for an increase in consumption of sugars such as glucose, by the respiratory pathways.

Nutritional research on carpogenesis naturally leads us to the problem of metabolic aspects of this phenomenon. Fragmentary data are known on this subject, but they do not actually permit the elaboration of a satisfactory general schema of the mechanisms involved in the release of fructification. The most complete works, to our knowledge, are those published by Wessels (1965) concerning Schizophyllum commune. The author shows that the metabolic activity of a culture of this fungus, tested by gas exchange is, first of all a "fermentative aerobic" type and then becomes a "respiratory" type when the initiation phase starts. Wessels thus established, for S. commune, a schema considered as having general value by many authors, such as Turian (1969), who had studied the metabolic aspects of the reproduction of various fungi. In fact, this metabolic evolution appears to be linked to the quantitative variations of different metabolites, particularly carbohydrates, firstly in the nutritional medium and then in the vegetative mycelium. Thus, Wessels showed that the fruit-bearing initiation of S. commune could only be produced if exogenous carbon and nitrogen sources were available. It will be seen that, after the initial stage, the nutritional needs are different in the case of this fungus. Studying the initiation of fructification of Coprinus lagopus, Stewart & Moore (1974) established that the quantities of reducing sugars and nitrogen (in the form of alpha-aminonitrogen) decreased in the nutritive medium from the beginning of the culture to the observation of the first carpophore buds. In the case of C. congregatus, Robert (1977b) showed that the fruit-bearing initiation occurred when glucose, in its quasi totality, was resorbed from the medium. One can still notice that the pH of the medium tends to become more basic when the primordia appear (as in C. lagopus, Stewart & Moore, 1974; C. congregatus, Robert, 1977b and Horrière,

1977).

In connection with nutritional and metabolic problems of fruit-bearing initiation, we can cite several cases of stimulation, by micro-organisms, of the fructification of superior fungi. Numerous observers have mentioned the preferential appearance of carpophores in mycelial areas contaminated by a Penicillium or by a bacterium. Salas & Hancock (1972) pointed out that P. oxalicum permits a spectacular increase of basidiocarps production in Mycena citricolor, when the two species are cultivated simultaneously. Authors attribute the Penicillium effects to one or more "basidiocarp stimulating substances" (BSS). The problem which arises by the eventual intervention of microorganisms in the release of fructification of Agaricus bisporus presents interest on the fundamental and on the application levels (cf. Couvy, 1973-74). As a matter of fact, cultivated Agaricus does not fructify in sterile conditions. Many authors (Park & Agnihotri, 1969; Eger, 1972 ...) considered that the micro-organisms from the earth which cover the cultures, have a positive effect on fructification.

b) Role of Humidity. This role is not well known. Many research workers consider that the fleshy "Macromycetes" fructification generally necessitates a high percentage of atmospheric humidity (Flammulina-Collybia-velutipes, according to Plunkett, 1956). But, on the contrary, an excessive atmospheric humidity can, in certain experimental conditions, prevent the appearance of primordia (Polyporus brumalis, according to Plunkett, 1956).

c) Role of Carbon Dioxide. An excess of atmospheric carbon dioxide generally inhibits fructification (Schizophyllum commune, Niederpruem, 1963). However, Polyporus brumalis fructifies under heavy levels of carbon dioxide (Plunkett, 1956). But as in the case of S. commune, in a majority of species (such as Agaricus bisporus, Tschierpe & Sinden, 1964, 1965) the presence of an excess of atmospheric carbon dioxide, for example at one percent, inhibits the formation of primordia. One can therefore understand the attention that mushroom cultivators must imperatively pay attention to the ventilation conditions of the culture.

d) Role of Temperature. This factor evidently intervenes in a different manner, according to the species, but equally for a given species, according to the development phases. Temperatures favouring preliminary growth of vegetative mycelia can differ to those permitting fruit-bearing initiation. Thus for three *Agaricus* (*A. bisporus*, *A. arvensis*, *A. sylvicola*, Couvy, 1973-74), incubation at 23 - 25°C is necessary to accomplish the vegetative phase and the optimal temperature, after which a decrease in temperature of 7 to 9°C is necessary to permit the release of fructification.

e) Role of Light. Illuminations are necessary to obtain the release of fructification of numerous species of "superior" and "inferior" fungi. This has been known for more than a century (Brefeld's observations in 1877, later confirmed by many other authors, e.g. Marsh, Taylor & Bassler, 1959; Durand, 1976 ...). The mycelia of such species stay sterile if they are kept in uninterrupted darkness. Nevertheless, according to Couvy (1973-74), light is not indispensable to the fruit-bearing induction of a great majority of *Agaricus* (*A. bisporus*, *A. arvensis*, *A. sylvicola*); this is equally true for many other species (various *Coprinus*, Brefeld, 1877; Buller, 1909 to 1934; *Lentinus tigrinus*, Schwantes, 1968-69 ...). The vegetative mycelia of such fungi produce in darkness either normal carpophores or, more often, abnormal primordia with thin stipes and reduced pilei, (certain *Coprinus*, *Lentinus tigrinus*, etc ...).

In summary, from the examples mentioned above, it is obvious that the release of Basidiomycetes fructification, from a mycelium genetically apt to fructify, is accompanied by notable metabolic variations (of which we do not know if they are "causes" or "consequences" of fruit-bearing initiation) and depends on variations characteristic to one or more chemical or physical environmental factors. In certain cases, an "antinomy" is observed between the vegetative growth and the initiation of reproduction: any factor favourizing one of these phases tends to prevent the accomplishment of the other. In fact, it is well known that this "equilibrium" does not only concern the mycological domain, other examples are also known in the vegetal and in the animal kingdoms.

2. The Morphogenetical Phase "Sensu Stricto"

As in the preceding phase of fruit-bearing initiation, this phase is dependent on various environmental factors, and also on endogenous factors.

Role of "Exogenous" Factors in Basidiocarp Formation. a) Nutritional and Metabolic Aspects. These aspects are too intimately imbricated to be distinguished here. To simplify this paper, only the species taken as examples earlier will be considered. As far as *Schizophyllum commune* is concerned, a Basidiomycete studied with precision, Wessels (1965) established that once the primordia are formed, the respiratory ratio superior to 1 until then, decreases and moves towards 1. The consumption of oxygen remains high as long as the exogenous glucose has not been totally exhausted and then falls sharply until a decrease during carpophore maturation. During primordia growth, if exogenous glucose is indispensable, this is not the case for an eventual exogenous nitrogen source: primordia needs can be satisfied by means of nitrogen compounds of vegetative mycelium. Subsequently, during the final phase of pilei formation, exogenous carbon or nitrogen sources are not necessary, but what is more, the supply of an exogenous carbon source obstructs pilei formation. Wesselsshowed that the pilei formation of carpophores perfecting their development depends on a feeble but continued supply of glucose, resulting from the hydrolysis of glucan membranes following the degradation of aborted primordia cell walls; in fact, on a vegetative thallus of *S. commune* or other Basidiomycetes, all the carpophore buds do not accomplish their development, only certain buds among these reach the terminal stage of maturity. The few data given above show the complexity of the problem. The progression of basidiocarp morphogenesis depends on complex nutritive and metabolic correlations, "nutritive medium - vegetative mycelium", "vegetative mycelium - primordia" and "dominating primordia - dominated primordia". On the other hand, basidiocarp morphogenesis is also a function of inter-relations between the stipe and the pileus: thus, the terminal phase of carpophor development of *C. congregatus* depends, according to Robert (1977a), on the transfer of carbohydrates and proteins from the stipe to the maturing pileus. The final morphogenesis of this Coprinus is not for all that autonomous because, according to Bret (1977a), the excised carpophores, in other words separated from the original mycelium, cannot reach a normal size even when the excision is done during the final hours of development. We can mention here Borriss's observations (1934a) according to whom glycogen, first of all concentrated in the lower and peripheral parts of the Coprinus lagopus stipes, would progressively "move" to the superior parts where it would finally be consumed during basidiospore formation.

Other examples could have been cited; certain histochemical works such as those of Komagata & Okunishi (1969) who, in the case of Coprinus kimurae, showed that the cytochrome-oxidase, succinic-dehydrogenase, acid and alkaline phosphatase activities were localized in the active growth zones of carpophores, essentially the superior part of the stipe and the edge of the pileus.

In practice, much work has to be done to understand the metabolism of superior fungus fructification, Agaricales and others. The few data just reported, which essentially concern carbohydrate metabolism in relation to basidiocarp morphogenesis, are a proof of this.

b) Role of Humidity. This factor is certainly one of the most important, but its action varies notably according to the species. Thus, Plunkett (1956) established that the carpophore buds of Flammulina (Collybia) velutipes do not develop normally unless the atmosphere is humid and sufficiently illuminated. Inversely, the development of Polyporus brumalis primordia, studied by the same author, is only normal if the air is a little humid and also under illumination. In the case of this latter species, the effects of a decreased illumination and a concomitant little transpiration in humid atmosphere, determines a relative increase in the final length of the stipe and a supression, practically total, of pileus development. Experiments of the same type as Plunkett's show that normal morphogenesis of basidiocarps is, in fact, a function of interactions of various environmental factors.

c) Role of Carbon Dioxide. The levels of atmospheric carbon dioxide equally condition the carpophore morphogenesis. An excess perturbs the development of the cultivated *Agaricus* primordia by delaying and then determining an excessive elongation of the stipe and a correlative pileus reduction, the last two anomalies being frequently linked. Schwantes (1968-69) deduced a similar need for appropriate ventilation to obtain normal development of the *Lentinus tigrinus* pileus. Plunkett (1956) ascertained that the reaction of *Collybia velutipes* buds to an excessive degree of carbon dioxide is similar to that of the cultivated *Agaricus* and *L. tigrinus*. The same author established that an excess of atmospheric carbon dioxide does not affect the morphogenesis of *Polyporus brumalis* primordia.

d) Role of Temperature. Few data have been consecrated to the influence of this factor on basidiocarp morphogenesis. Brefeld (1877), mentioned that an increase in temperature leads to a relatively more normal development of *Coprinus stercorarius* carpophores, produced on mycelia obtained in darkness and which, finally, would have presented a thin stipe and a very reduced pileus. In the case of *C. congregatus*, the complete development of carpophores is possible in a large range of temperatures, but it is much slower at low temperatures; approximately five days at 25°C are necessary, 8 - 12 days at 15°C and 15 - 22 days at 10°C (Robert & Manachere, 1971-73).

e) Role of Light. The photomorphogenesis of all types of fungi has been the object of innumerable observations, without the precise role of light having been established (cf. Marsh, Taylor & Bassler, 1959; Durand, 1976 ...). Rare are the species where it has been possible to establish that mycelia produce absolutely normal carpophores in darkness. More currently, non-lighted mycelia either stay sterile (e.g. *C. congregatus*, already mentioned, Manachère, 1970) or produce primordia with thin stipes terminated by very reduced pilei, as opposed to the normal (e.g. *Coprinus lagopus*, Borriss, 1934a, b, and numerous other species ...); sometimes the "etiolated" carpophores, in the absence of light, carry abnormal lateral branches which appear just like carpophore buds (Lentinus tigrinus, Schwantes, 1968-69; Asterophora parasitica, Viot, 1970). In all these cases, only illuminations of suitable lengths and intensities, provided at the particular stages of development, can permit normal morphogenesis.

Among the photo-sensitive Agaricales, C. congregatus - studied in our laboratory at Lyon University - appears to be the most complex species (Manachère, 1967; 1970). At a temperature of 25°C it was established that light, first of all, determined the progress of morphogenesis during the course of a first "photo-stimulated" phase, until a stage which we call "sensitiveness to darkness", corresponding to the minus-36 hour stage according to our earlier definitions. As from this stage, the light, always at 25°C, prevents stipe elongation and the evolution of basids towards normal sporogenesis. A minimum of 4 hours darkness (12 are suitable) is necessary to achieve morphogenesis (Manachère, 1966a). This darkening must be followed by an illumination to accomplish a second "photo-stimulated" phase which permits normal maturation of the majority of basidiospores, with the spreading of the pileus followed by a progressive pileus autolysis. A premature interruption of illumination during the accomplishment of the first photo-stimulated phase leads to the production of thin stipes and reduced pilei primordia. An analogous interruption during the course of the second photo-stimulated phase does not prevent stipe elongation but does not permit the pileus opening and strongly reduces basidiospore production. It would seem that light is particularly necessary to normal pileus development until sporogenesis and that, when these illumination conditions become unfavourable, only stipe elongation can be achieved. The photo-inhibited phase of primordia development in C. congregatus is of particular interest. We have established that the end of the darkness period permitting its accomplishment, in other words the beginning of the photoperiod which permits the realization of the second photo-stimulated phase, determines the moment for the end of carpophore development, more or less 24 hours after. Numerous fungal species belonging to the most different systematic groups reach their development and project their spores at fixed hours in nature. It would seem that the determinisms of these "programmed" phenomena are of the type known in the case of C. congregatus. On the other hand, it was established (Robert, 1971) that the darkness period (from 4 - 12 hours) necessary and sufficiant for the accomplishment of the photo-inhibited phase under a uniform temperature of 25°C, can be "replaced" by a fall in temperature of 15°C, the cultures therefore staying under uninterrupted illumination. The following question remains unanswered: does the darkening and fall in temperature, in these cases, determine the accomplishment of identical, or different, physiological sequences, even when leading to the same results, basidiocarps maturation? Whatever the outcome, these results, after those of Plunkett (1956) already noted concerning Flammulina (Collybia) velutipes and Polyporus brumalis, confirm that, in the future, research on the physiology of fructification should even more consider the interactions of various intervening factors.

It is necessary to mention here that the carpophores of numerous Basidiomycetes are characterized by a positive phototropism, at least at the juvenile stage (some Coprinus, Buller, 1909, 1930; Coprinus lagopus, Borriss, 1934; Polyporus brumalis, Plunkett, 1961; Coprinus congregatus, unpublished own observations). Moreover, at the end of development (between the minus 24-hours stage and minus 12-hours stage in the case of C. congregatus), the carpophore becomes exclusively subject to a negative geotropism. The tip of the stipe therefore tends towards the vertical in such a manner that, at the moment of sporulation, the sporogenous lamellae of Agaricales (and the hymenial pores of Boletales or Polyporales) will most often assure the basidiospores the best chance of being projected and then dispersed into the atmosphere.

The very numerous works intended to determine the influence of various luminous radiations on the different stages of fungal reproduction are not covered here. It will only be recalled that near ultra-violet and visible radiations of short-wave lengths seem, above all, to be implicated (cf. Durand, 1976).

"Endogenous" Mechanisms Regularizing Basidiocarp Formation. Many of the experiments mentioned earlier show that, when a factor prevents the pileus development, an "etiolation" of the stipe frequently follows. In fact, complex correlations regularize the respective developments of the two distinct parts of basidiocarps. Hagimoto & Konishi (1959, 1960), Hagimoto (1963a, b) and Gruen (1963) showed that one or more growth substances elaborated at the level of hymenial lamellae probably control the elongation of carpophore stipes in cultivated *Agaricus*. Therefore, the unilateral elimination of lamellae from a young carpophore leads to a preferential elongation at the side where the lamellae were conserved, and a curvature follows. Similar observations have been made in the case of *C. congregatus* by Bret (1977b). The substances responsible have not yet been recognized. It would seem that the relationships are one-way, the pilei controlling the stipes elongation. This is not the case, since the stipes evidently assure the translocation of various nutrients from the vegetative mycelium to the different parts of basidiocarps, and we have already mentioned that the latter can remain dependent on the original thallus until the final hours of development. Going from the information given above and by the analogy of our knowledge on the intervention of growth substances in tropisms of superior plants, it would seem reasonable to think that the positive phototropisms and negative geotropisms characterizing certain phases of basidiocarp morphogenesis could be determined by hormonal substances, emitted asymetrically by the hymenial lamellae of carpophores and determining the curvature of stipes by action on the elongation of the superior cells of latter.

At the mycelia level, other complex correlations exist. They show that the thalli constitute physiological unities. Thus, a culture of *C. congregatus* maintained at a temperature of 25°C under alternating periods of 12 hours in light and 12 hours in darkness, produces mature carpophores according to a characteristic rhythm: four or five days more often separate the beginning of photo-periods where the carpophores at the zero hour stage of successive flushes of fructification can be observed (Manachère, 1966b, 1971). An analogous rhythm is known in the case of the cultivated *Aga-*

ricus (Cooke & Flegg, 1962, 1965; Flegg, 1967) and in that of the Gasteromycete Sphaerobolus stellatus (Friederichsen & Engel, 1957, 1960; Alasoadura, 1963). In the case of these two fungi, as in the case of Coprinus, the elimination of primordia leads to an acceleration of carpophore development of the following flushes, the acceleration being even more important when the eliminated buds are younger. All this happens as if the buds of a fructification flush, during their accomplishment, exert an inhibition on other primordia destined to give carpophores of ulterior flushes. The inhibitory mechanism is actually unknown. One can postulate either nutrient diversion by the predominant primordia or the emission of inhibitory substances. In the particular case of C. congregatus, we have also shown that carpophore elimination, not at the primordial stage but at maturity, can reduce the accomplishment of ulterior flushes and even prevent them if the carpophores at the zero-hour stage of various successive flushes are eliminated (Manachère, 1966b). All this happens as if, once they have reached their maturity, the carpophores bring (or restore) to the original thallus, one or more substances stimulating ulterior fructification (Manachère, 1976a, b). We are currently searching to define more precisely the mechanisms regularizing the rhythm evoked. One should remember that the mastery of this periodic phenomenon is of constant worry to the producers of the cultivated Agaricus, who, for example, prefer an abundant production during the week or the day before holidays and, a reduced production during the days when man-power is little available, such as week-ends and bank holidays!

The carpophores of C. congregatus therefore exert, successively, on the producting mycelium, an inhibitory effect and then a stimulating effect according to whether they are at the juvenile or mature stage. We found it useful to search if the basidiocarps of this species also presented different potentialities when they were used as "inoculum" to obtain new cultures. Bret (1973, 1975) has confirmed that this is so: when the primordia of different ages are deposited on nutritive media, the primary viable buds essentially develop from inoculum-carpophores if they have not reached their maturity at the time of sampling (we call this direct regeneration); inversely, when the inoculum-carpophores are sampled at maturity, the primary regenerated primordia principally develop on the vegetative mycelia developed from inoculum-carpophores (indirect regeneration). Analogous results were obtained in the case of Asterophora parasitica (Viot, 1974). Concerning the "fruiting-stimulating" properties of various species of basidiocarps, see also Eger (1965a, b, 1968). It is noticeable that the substances involved seem non-specific.

CONCLUSIONS

During the course of this paper a limited range of examples was taken among a very large number of past and present data. To simplify them, they were classified under various headings but it is evident that the limits between these various headings are arbitrary, which proves, from our point of view, how desirable it is that the researchers concerned directly with one of them avoid an excessive and injurious specialisation. Global knowledge of the various problems is actually indispensable to understand each particular result and to integrate it into the general framework of the physiology of basidiocarp development.

REFERENCES

Alasoadura, S. O. (1963). Fruiting in Sphaerobolus with special reference to light. Ann. Bot. (G.B.) 27, 123-145.

Bille-Hansen, E. (1953a). Fructification of a coprophilous Coprinus on synthetic medium. Physiol. Plantarum 6, 523-528.

Bille-Hansen, E. (1953b). Fructification of three coprophilous species of Coprinus using glucose, sucrose and maltose as carbon source. Bot. Tidsskrift 50, 81-85.

Borriss, H. (1934a). Beiträge zur Wachstums- und Entwicklungsphysiologie der Fruchtkörper von Coprinus lagopus. Planta 22, 28-69.

Borriss, H. (1934b). Über den Einfluß äußerer Faktoren auf Wachstum und Entwicklung der Fruchtkörper von Coprinus lagopus. Planta 22, 644-684.

Brefeld, O. (1877). Botanische Untersuchungen über Schimmelpilze, III. Arthur Felix, Leipzig.

Bret, J. P. (1973). Variations du potentiel de régénération des carpophores de Coprinus congregatus Bull. ex Fr. en fonction de leur stade de développement. C. R. Acad. Sci., Série D (Paris) 276, 1847-1850.

Bret, J. P. (1975). Etude comparative des potentiels de régénération des carpophores de Coprinus congregatus Bull. ex Fr. développés dans différentes conditions de température ou d'éclairement. Biologia Plantarum 17, 347-351.

Bret, J. P. (1977a). Respective role of cap and mycelium on stipe elongation of Coprinus congregatus. Trans. Brit. Mycol. Soc. 68 (in press).

Bret, J. P. (1977b). Contrôle de l'élongation du stipe par le chapeau dans les carpophores de Coprinus congregatus. C. R. 102 è Congrès National Soc. Savantes, Limoges, Avril 1977 (in press).

Buller, A. H. R. (1909). Researches on fungi, vol. I; (1922) ibid., vol. II; (1924) ibid., vol. III; (1931) ibid., vol. IV; (1933) ibid., vol. V; (1934) ibid., vol. VI; Nouv. édit. (1958) Hafner Publishing Co., New York.

Burnett, J. H. (1968). Fundamentals of mycology. Edward Arnolds (Publishers) Ltd., London.

Cooke, D. & Flegg, P. B. (1962). The relation between yield of the cultivated mushroom and stage of maturity at picking. J. hort. Sci. 37, 167-174.

Cooke, D. & Flegg, P. B. (1965). The effect of stage of maturity at picking on the flushing of crops of the cultivated mushroom. J. hort. Sci. 40, 207-212.

Couvy, J. (1973-1974). Les facteurs de fructification des Agaricales et plus particulièrement de l'Agaricus bisporus (Lange) Sing. Le Botaniste 56, 103-128.

Durand, R. (1976). Influence des radiations lumineuses sur les processus de reproduction des champignons; hypothèses sur l'identité des photorécepteurs. Revue bibliographique. Mycopathologia 60, 3-16.

Eger, G. (1965a). Untersuchungen über die Bildung und Regeneration von Fruchtkörpern bei Hutpilzen. I. Pleurotus florida. Arch. Mikrobiol. 50, 343-356.

Eger, G. (1965b). Untersuchungen über die Bildung und Regeneration von Fruchtkörpern bei Hutpilzen. III. Flammulina velutipes Curt. ex Fr. und Agaricus bisporus (Lge) Sing. Arch. Mikrobiol. 52, 282-290.

Eger, G. (1968). Untersuchung zur Stimulation der Fruchtkörperbildung bei einem Pleurotus (aus) Florida. Theor. appl. Genet. (All.) 38, 23-27.

Eger, G. (1972). Experiments and comments on the action of bacteria on sporophore initiation in Agaricus bisporus. Mushroom Science 8, 719-725.

Esser, K. (1977). Genetic aspects of differentiation. Vienna Symposium of the Federation of European Microbiological Societies. March 28 - April 1, 1977.

Flegg, P. B. (1967). The effect of harvesting procedures on yield, stage of maturity at harvest and number of pickings of cultivated mushroom. J. hort. Sci. 42, 139-148.

Friederichsen, I. & Engel, H. (1957). Beiträge zur Kenntnis des Abschußrhythmus und des Farbstoffs von Sphaerobolus stellatus (Thode) Pers. Planta 49, 578-587.

Friederichsen, I. & Engel, H. (1960). Der Abschußrhythmus des Fruchtkörper von Sphaerobolus stellatus (Thode) Pers. Planta 55, 313-326.

Gruen, H. E. (1963). Endogenous growth regulation in carpophores of Agaricus bisporus. Plant Physiol. (USA) 38, 652-666.

Hagimoto, H. (1963a). Studies on the growth of fruit-body of fungi. IV. The growth of the fruit-body of Agaricus bisporus and the economy of the mushroom growth hormone. Bot. Mag. Tokyo 76, 256-263.

Hagimoto, H. (1963b). The mushroom growth hormone and geotropic response of fruit-body in Agaricus bisporus. Bot. Mag. Tokyo 76, 363-365.

Hagimoto, H. & Konishi, M. (1959). Studies on the growth of fruit-body of fungi. I. Existence of a hormone active to the growth of fruit-body in Agaricus bisporus (Lange) Sing. Bot. Mag. Tokyo 72, 359-366.

Hagimoto, H. & Konishi, M. (1960). Studies on the growth of fruit-body of fungi. II. Activity and stability of the growth hormone in the fruit-body of Agaricus bisporus

(Lange) Sing. Bot. Mag. Tokyo 73, 283-287.

Hawker, L. E. (1942). The effect of vitamin B1 on the concentration of glucose optimal for the fruiting of certain fungi. Ann. Bot. (G.B.) 6, 631-636.

Hawker, L. E. (1966). Environmental influences on reproduction - in Ainsworth, G. C. & Sussmann, A. S. 1966 - The fungi, an advanced treatise, vol. II: The fungal organism. 435-469. Academic Press, New York & London.

Horrière, F. (1976). Influence sur la fructification de Coprinus congregatus Bull. ex Fr. des concentrations en glucose et en asparagine, et de l'équilibre carbone-azote du milieu. Mycopathologia 58, 81-89.

Horrière, F. (1977). Recherches physiologiques sur la fructification d'un Basidiomycète, Coprinus congregatus Bull. ex Fr.: influence de quelques facteurs nutritionnels. Thèse Ingénieur-Docteur, Univ. Lyon I.

Kühner, R. (1926). Contribution à l'étude des Hyménomycètes et spécialement des Agaricacées. Thèse Sciences, Paris.

Komagata, K. & Okunishi, M. (1969). Histochemical studies on the development of carpophore of Coprinus kimurae. J. Gen. Appl. Microbiol. 15, 231-237.

Manachère, G. (1966a). Asservissement à une nyctipériode de la fructification de Coprinus congregatus Bull. ex Fr. en cultures éclairées. C. R. Acad. Sci., Série D (Paris), 262, 482-484.

Manachère, G. (1966b). Périodicité de la fructification de Coprinus congregatus Bull. ex Fr. en cultures éclairées. C. R. Acad. Sci., Série D (Paris) 262, 2711-2714.

Manachère, G. (1967). Déroulement de la fructification de Coprinus congregatus Bull. ex Fr. en relation avec les conditions d'éclairement. Bull. Soc. Mycol. Fr. 85, 257-285.

Manachère, G. (1970). Recherches physiologiques sur la fructification de Coprinus congregatus Bull. ex Fr.: action de la lumière; rythme de production de carpophores. Ann. Sci. Nat. Bot. et Biol. Vég. (Paris) 11, 1-95.

Manachère, G. (1971). Research on the fruiting rhythm of a basidiomycete mushroom Coprinus congregatus Bull. ex Fr. J. interdiscipl. Cycle Res. 2, 199-209.
Manachère, G. (1976a). Substance (s) produite (s) par les carpophores mûrs de Coprinus congregatus et régularisatrice (s) du rythme de fructification du champignon; délais nécessaires à leur libération. Rev. Mycol. (Fr.) 40, 329-339.
Manachère, G. (1976b). Changes in fruiting rhythm of Coprinus congregatus by addition of exogenous mature sporophores on the surface of cultures. Physiol. Plantarum (in press).
Manachère, G., Robert, J. C., Bret, J. P. & Hugueney, R. (1974). Recherches physiologiques sur la fructification de Coprinus congregatus Bull. ex Fr. Perspectives nouvelles. Bull. Soc. Linnéenne Lyon, 43ème année, n° spécial, travaux mycologiques dédiés à R. Kühner, 255-283.
Marsh, P. B., Taylor, E. E. & Bassler, L. M. (1959). A guide to the literature on certain effect of light on fungi: reproduction, morphology, pigmentation and phototropic phenomena. Plant Dis. Rep. (USA) suppl. 261, 251-312.
Matthews, T. R. & Niederpruem, D. J. (1972). Differentiation in Coprinus lagopus. I. Control of fruiting and cytology of initial events. Arch. Mikrobiol. 87, 257-268.
Niederpruem, D. J. (1963). Role of carbon dioxide in the control of fruiting of Schizophyllum commune. J. Bact. 85, 1300-1308.
Park, J. Y. & Agnihotri, V. P. (1969). Bacterial metabolites trigger sporophore formation in Agaricus bisporus. Nature 222, 984.
Plunkett, B. E. (1956). The influence of factors of the aeration complex and light upon fruit-body form in pure cultures of an Agaric and a Polypore. Ann. Bot. (G. B.) 20, 563-586.
Plunkett, B. E. (1961). The change of tropism in Polyporus brumalis stipes and the effect of directional stimuli on pileus differentiation. Ann. Bot. (G.B.) 25, 206-223.

Reijnders, A. F. M. (1963). Les problèmes du développement des carpophores des Agaricales et de quelques groupes voisins. Uitgeverij Dr. W. Jung, La Haye.

Robert, J. C. (1971). Effet favorable d'une période froide sur la maturation de carpophores de *Coprinus congregatus* Bull. ex Fr. inhibés par un éclairement continu. C. R. Acad. Sci., Série D (Paris) 273, 154-157.

Robert, J. C. (1977a). Fruiting of *Coprinus congregatus*: biochemical changes in fruit-bodies during morphogenesis. Trans. Brit. Mycol. Soc. 68 (in press).

Robert, J. C. (1977b). Fruiting of *Coprinus congregatus*: relationship to biochemical changes in the whole culture. Trans. Brit. Mycol. Soc. 68 (in press).

Robert, J. C. & Manachère, G. (1971), (paru 1973). Influence de la température sur le rythme de fructification de *Coprinus congregatus* Bull. ex Fr.: remarques préliminaires. C. R. 96ème Congrès National Soc. Savantes Toulouse, Avril 1971, paru en 1973, 3, 119-129.

Salas, J. A. & Hancock, J. G. (1972). Production of the perfect stage of *Mycena citricolor* (Berk. and Curt.) Sacc. 41, 213-234.

Schwantes, H. O. (1968 - 1969). Wirkung unterschiedlicher Stickstoffkonzentrationen und -verbindungen auf Wachstum und Fruchtkörperbildung von Pilzen. Mushroom Science 7, 257-272.

Stewart, G. R. & Moore, D. (1974). The activity of glutamate dehydrogenases during mycelia growth and sporophore development in *Coprinus lagopus* (sensu Lewis). J. Gen. Microbiol. 83, 73-81.

Tschierpe, H. J. & Sinden, J. W. (1964). Weitere Untersuchungen über die Bedeutung von Kohlendioxid für die Fruktifikation des Kulturchampignons, *Agaricus campestris* var. bisporus (L.) Lge. Arch. Mikrobiol. 49, 405-425.

Tschierpe, H. J. & Sinden, J. W. (1965). Über leicht flüchtige Produkte des aeroben und anaeroben Stoffwechsels des Kulturchampignons, *Agaricus campestris* var. bisporus (L.) Lge. Arch. Mikrobiol. 52, 231-241.

Turian, G. (1969). Différenciation fongique. Masson, Paris.

Uno, I. & Ishikawa, T. (1971). Chemical and genetical control of induction of monocaryotic fruiting-bodies in Coprinus macrorhizus. Molec. gen. Genetics 113, 228-239.

Uno, I. & Ishikawa, T. (1973). Purification and identification of the fruiting inducing substances in Coprinus macrorhizus. J. Bact. 113, 1240-1248.

Uno, I. & Ishikawa, T. (1974). Effect of glucose on the fruiting-body formation and adenosine 2', 5'-cyclic monophosphate levels in Coprinus macrorhizus. J. Bact. 120, 96-100.

Viot, M. (1970). Mise en évidence de corrélations de croissance au cours du développement des carpophores d' Asterophora parasitica (Bull. ex Fr.) Sing. C. R. Acad. Sci. Série D (Paris) 270, 790-792.

Viot, M. (1974). Etude comparative du pouvoir de régénération de carpophores d'Asterophora parasitica (Bull. ex Fr.) Sing. développés à la lumière ou à l'obscurité. C. R. Acad. Sci., Série D (Paris) 278, 265-268.

Volz, P. A. & Beneke, E. S. (1969). Nutritional regulation of basidiocarp formation and mycelial growth of Agaricales. Mycopathologia et Mycologia applicata 37, 225-253.

Wessels, J. G. H. (1965). Morphogenesis and biochemical processes in Schizophyllum commune Fr. Wentia 13, 1-113.

Zur Lippe, T. & Nesemann, G. (1959). Über die Fruchtkörperbildung von Merulius lacrymans domesticus Falk. Arch. Mikrobiol. 34, 132-148.

GENETIC ASPECTS OF DIFFERENTIATION IN FUNGI

K. Esser, U. Stahl und F. Meinhardt

Differentiation is one of the most fascinating phenomena in biology, which at present is studied in many organisms under various aspects. At the beginning of this decade the experimental study of differentiation was brought to a broader scientific attention by Spemann's classical studies on embryogenesis in animals. Plants (and certainly fungi) were not considered in those days as objects for studies of this kind. This explains why in text books or scientific dictionaries (e.g. Rieger et al., 1976) one finds differentiation mostly considered from the zoological viewpoint.

There is a useful distinction between two phases of differentiation: intracellular or cytodifferentiation, and intercellular differentiation or histogenesis. Cytodifferentiation is thought to be the prerequisite for histogenesis and the ensemble of these processes leads to morphogenesis. In fungi which at present are in fashion for morphogenetic studies, because of their simple organisation and easy handling, these definitions cannot be applied in any strict sense, because there are no genuine "tissues", and even, in many cases, there is no articulation of thalli in real cells. Therefore it has become common to use both terms (differentiation and morphogenesis) simultanously for the same phenomena.

The earlier studies of differentiation in fungi were mostly initiated with the object of understanding the sexual reactions. Morphogenesis was more or less noticed as a byproduct. I may recall as examples the analyses of sexual hormones in the Mucoraceae and in the water mold Achlya initiated by Burgeff and John Raper respectively. However, particularly in the Myxomycetes differentiation was studied for itself: we can recall the first studies of Kenneth Raper and Bonner leading eventually to the detection of cyclic AMP as a "morphogenetic substance" (for literature see Jacobson & Lodisch, 1975; Esser & Kuenen, 1967; Gerisch et al., 1974; Scott, 1976).

However, all these and other comprehensive studies have rarely taken the genetic aspect of differentiation into account, mainly because although fungi are very excellent for physiological experiments, the ones mentioned are particularly inaccessible to a genetic analysis.

Already in the thirties, well aware that these morphological and physiological results needed a genetic complement, Max Hartmann had encouraged his collaborators to perform genetic studies with fungi, but unfortunately they worked exclusively under his dogma of relative sexuality, without recognizing the chances of studying morphogenesis (Hartmann, 1943).

More or less following the "spirit of the time" we started some 20 years ago our genetic research with the ascomycete Sordaria macrospora (Esser & Straub, 1958). The results summarized in Fig. 1 allow the following conclusions:

1.) There is a sequence of at least 14 genes which control the life cycle, i. e. the morphogenesis, of this fungus. This was a new finding because sexual differentiation was thought to be controlled either by sex chromosomes or by single genes like mating type factors.

2.) The possibility that all these deficiency mutants complemented in the heterokaryon and restored the wild phenotype, finally ruled out Hartmann's concept of relative sexuality.

In the following years this concept of a polygenic control of sexual morphogenesis was confirmed in some other fungi, like Neurospora, Podospora, Phycomyces (ref. in Esser & Kuenen, 1967; Tatum, 1973). Since the perithecia (the fruit bodies of Sordaria) and their initials are very small and only accessible to microscopic observations, we switched to higher basidiomycetes (especially to mushrooms) where no sex organs are present. Here, morphogenesis is less complicated because it consists mainly in a transition from mycelial growth to stromatic growth. In addition the basidiomycetes have been studied genetically very thoroughly, at least so far as the ma-

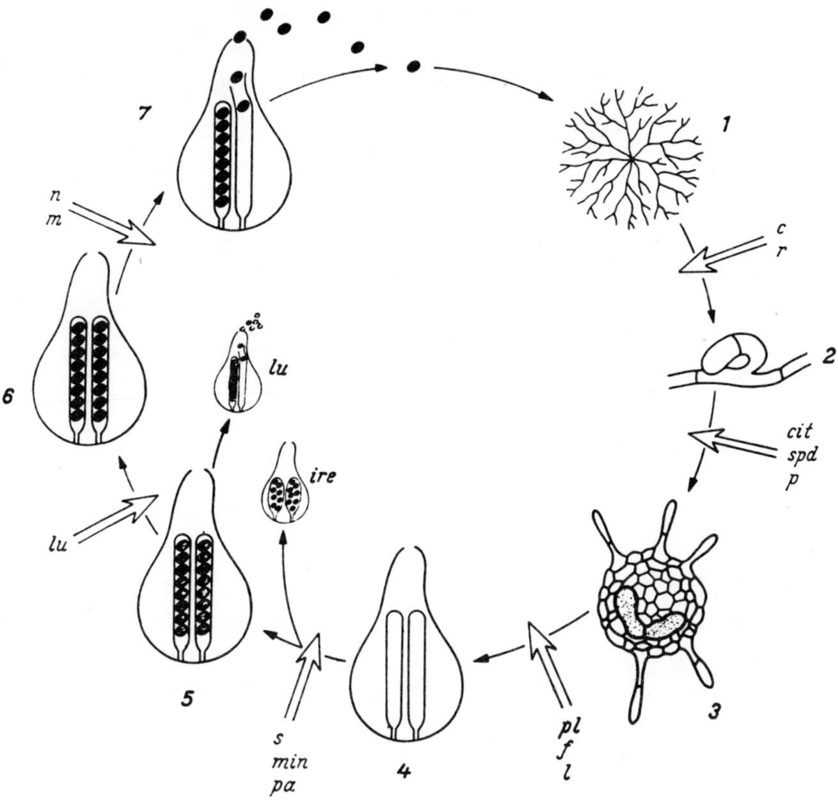

Fig. 1. Life cycle of Sordaria macrospora. The significant differentiation steps are: mycelium formation (1), differentiation of the ascogonium (2), transformation of the ascogonium into a protoperithecium (3), dikaryotic phase, karyogamy, and ascus formation (4), meiosis and spore formation (5), maturation of the ascospores (6), discharge of the ascospores (7). Male sex organs are not produced in this apandrous fungus. The arrows with the gene symbols show at which stage of differentiation particular mutations block the developmental sequence. Further explanation in text. (From Esser & Straub, 1958).

ting types and the incompatibility/compatibility reaction is concerned (for lit. see Raper, 1966). As may be seen from Fig. 2 (upper left):

1.) The incompatibility factors A and B control the va-

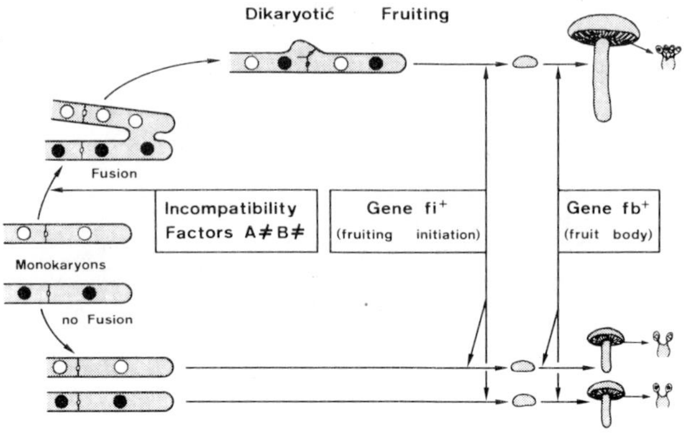

Fig. 2. Genetic control of fruit body formation in *Agrocybe aegerita* (Agaricales). Both dikaryotic fruiting (sexual cycle) and monokaryotic fruiting (asexual cycle) start from monokaryotic mycelia originating from single basidiospores. The genetic difference of nuclei due to different incompatibility factors is marked by black and white color. Monokaryotic fruit bodies reach only about 1/3 of the size of the dikaryotic, but their basidia show only a difference in spore number not in size. For further explanations see text. (Adapted from Esser & Meinhardt, 1977).

rious morphogenetic steps from the fusion of monokaryons (plasmogamy) to the establishment of a dikaryon. The latter guarantees (by the formation of clamp connections) the equal distribution of both parental types of nuclei during mycelial growth and fruit body differentiation until finally karyogamy and subsequently meiosis takes place in the young basidia.

2.) No genetic factors were found controlling the initiation and differentiation of fruit bodies. For this essential morphogenetic sequence in some mushrooms only the environmental conditions were known.

We arrived at an understanding of the genetic control of fruit body morphogenesis in dikaryotic mycelia by way of the detour of a genetic analysis of the so-called monokaryotic fruiting.

Monokaryotic fruiting is the ability of a monokaryon to form fruitbody or at least fruitbody-like structures. This phenomenon has been known for some time (Maire, 1902) and has been described in many Poriales and Agaricales (see list in Stahl & Esser, 1976), but it was never analysed in detail morphologically and genetically.

We have studied monokaryotic fruiting, and its implications for dikaryotic fruiting, in two not very closely-related higher basidiomycetes: Polyporus ciliatus (Polyporales) and Agrocybe aegerita (Agaricales) (Esser & Stahl, 1973; 1975; Esser et al., 1974; Stahl & Esser, 1976; Esser & Hoffmann, 1977; Esser & Meinhardt, 1977).

It could be shown that in some of the monokaryotic fruit bodies presented in Fig. 3 and 4, basidia with fertile basidiospores were formed. Cytological observations revealed that in these monokaryotic basidia neither a karyogamy nor a meiosis occurred, but only a mitosis. This also explained the fact that the monokaryotic basidia formed two spores only, instead of the four spores produced by dikaryotic basidia (see Fig. 2).

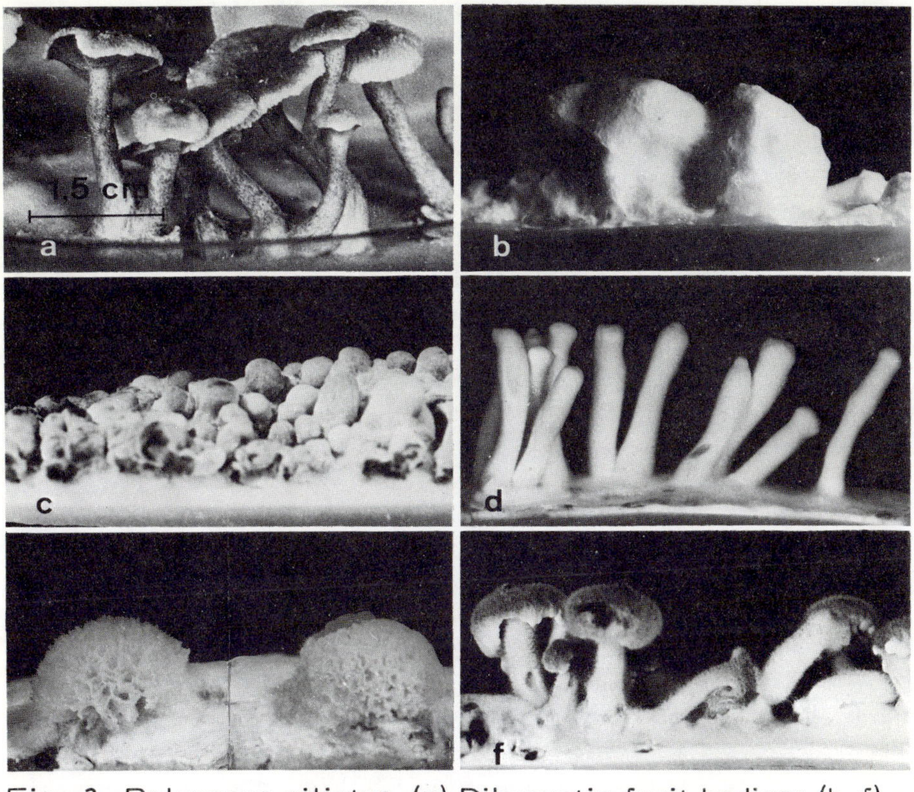

Fig. 3. Polyporus ciliatus. (a) Dikaryotic fruit bodies; (b-f) various types of monokaryotic fruit bodies (Stahl & Esser, 1976)

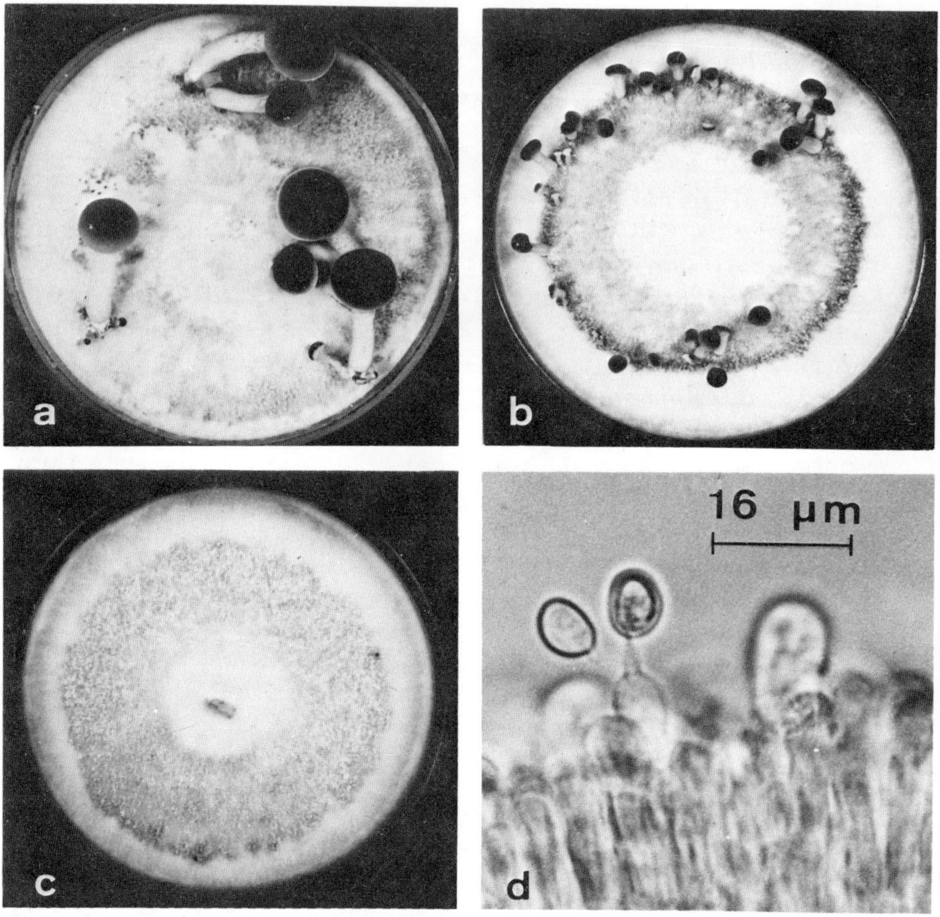

Fig. 4. Agrocybe aegerita. (a) Dikaryotic fruit bodies; (b) monokaryotic fruit bodies; fi^+ fb^+ type; (c) monokaryotic fruiting initials (fi^+ fb); (d) two-spored basidia from a monokaryotic fruit body. Further explanation see text. (Adapted from Esser et al., 1974).

Despite the great versatility of monokaryotic fruit bodies in both species (see Fig. 3 and 4), a common principle in the genetic control was found for fungi, which may be taken from the scheme of Fig. 2:

1.) There is one principal gene, called fi^+, which initiates the morphogenesis of fruit bodies, i.e. it is responsible for crossing over the threshold from merely mycelial growth to differentiation growth.

2.) At least one more gene (fb^+) is afterwards effective

in determining the shape of the fruit body.

3.) The subsequent action of both genes is needed for the formation of fertile fruit bodies (fi^+ fb^+). The ineffective allele fb of the second gene will block the differentiation after the formation of fruit body initials (fi^+ fb). The ineffective allele fi will however block before this step, independently of the constitution of the fb locus (fi fb^+ or fi fb).

4.) It was also found that these genes, detected in monokaryotic fruiters, are equally effective and necessary for dikaryotic fruiting. In the dikaryon one dose of each of both genes is needed, therefore both the cis (fi^+ fb^+ × fi fb) and the trans configuration (fi^+ fb × fi fb^+) will lead to fertile fruit bodies.

Since work still in progress in Schizophyllum commune (Esser, Saleh & Meinhardt unpubl.), the genetically best-known higher basidiomycete gave very strong indications of a similar genetic control as in P. ciliatus and A. aegerita, we dare to generalize our findings as follows:

As in Sordaria macrospora there is a straight genetic control for the initiation of differentiation, started by a single gene. The gene fi^+ initiates monokaryotic and dikaryotic fruiting. For the completion of fruit bodies at least one more gene (fb^+) is required.

Certainly these results need support by biochemical data. But even without this kind of experiment we think that they are of general importance for both basic and applied research as compiled in the following manner:

GENETIC CONTROL OF MORPHOGENESIS WITH RESPECT TO FRUIT BODY FORMATION IN HIGHER BASIDIOMYCETES

basic research	applied research
A single gene (fi^+) switches on morphogenesis by initiating the change from undirected mycelial growth to organized stromatic growth.	Concerted breeding is facilitated. It is now known that fruit body formation is not dependent only on external conditions. Sterile strains become fertile when crossed with fi^+ strains.

Other genes act subsequently to determine fruit body morphology.

Mutation of $\underline{fi}^{+}$ to its allele fi leads to an imperfect fungus.

This can be made use of in Biotechnology:

1. Simple procedures: production of mushrooms in recycling procedures.
2. Complex procedures: metabolites occurring only in fruit bodies may be produced in fermenters.

Summary. The genetic analysis of differentiation in higher fungi has shown that in fungi of different genera thoroughly analysed there is a single gene which initiates fruit body production and the onset of the sexual cycle. It is followed by other genes controlling the subsequent steps of morphogenesis. The knowledge of these genes together with other controlling genetic elements facilitates concerted breeding of fungi for applied purposes, especially for production of food and feed, for recycling, and also for fermentation objectives because many metabolites are only formed in fruit bodies or fruitbody-like structures.

References

Esser, K. & Hoffmann, P. (1977). Genetic basis for speciation in higher basidiomycetes with special references to the genus Polyporus. In: H. Clémencon (ed.): Species concept in hymenomycetes. J. Cramer, Vaduz (in press).

Esser, K. & Kuenen, R. (1967). Genetics of Fungi. Springer Berlin - Heidelberg - New York.

Esser, K. & Meinhardt, F. (1977). A common genetic control of dikaryotic and monokaryotic fruiting in the basidiomycete Agrocybe aegerita. (in press).

Esser, K. & Stahl, U. (1973). Monokaryotic fruiting in the basidiomycete Polyporus and its suppression by incompatibility factors. Nature 244, 304-305.

Esser, K. & Stahl, U. (1975). A genetic correlation between dikaryotic and monokaryotic fruiting in basidiomycetes.

Proc. 1st Intersect. Congr. IAMS Japan, Tokyo, 1974. Vol. 1, 294-300.

Esser, K. & Straub, J. (1958). Genetische Untersuchungen an *Sordaria macrospora* Auersw., Kompensation und Induktion bei genbedingten Entwicklungsdefekten. Z. Vererbungsl. 89, 729-746.

Esser, K., Semerdzieva, M. & Stahl, U. (1974). Genetische Untersuchungen an dem Basidiomyceten *Agrocybe aegerita*. I. Eine Korrelation zwischen dem Zeitpunkt der Fruchtkörperbildung und monokaryotischem Fruchten und ihre Bedeutung für Züchtung und Morphogenese. Theor. Appl. Genet. 45, 77-85.

Gerisch, G., Malchow, D. & Hess, B. (1974). Cell communication and cyclic AMP regulation during aggregation of the slime mold *Dictyostelium discoideum*. 25. Mosbacher Coll. 1974: Biochemistry of sensory functions. pp. 279-298. Springer, Berlin.

Hartmann, M. (1956). Die Sexualität. Fischer, Stuttgart.

Jacobson, A. & Lodisch, H. F. (1975). Genetic control of development of the cellular slime mold *Dictyostelium discoideum*. Ann. Rev. Genet. 9, 145-185.

Maire, R. (1902). Recherches cytologiques et taxonomiques sur les Basidiomycètes. Thèse, Paris.

Raper, J. R. (1966). Genetics of sexuality in higher fungi. Ronald Press New York.

Rieger, R., Michaelis, A. & Green, M. M. (1976). Glossary of genetics and cytogenetics. Springer, Berlin - Heidelberg - New York.

Scott, W. A. (1976). Biochemical genetics of morphogenesis in *Neurospora*. Ann. Rev. Microbiol. 30, 85-104.

Stahl, U. & Esser, K. (1976). Genetics of fruit body production in higher basidiomycetes. I. Monokaryotic fruiting and its correlation with dikaryotic fruiting in *Polyporus ciliatus*. Molec. gen. Genet. 148, 183-197.

Tatum, E. L. (1973). Biochemische Genetik der Morphogenese bei *Neurospora*. Naturwiss. Rdsch. 26, 140-142.

RIBONUCLEASES IN THE DIFFERENTIATION OF FUNGI

G. S. Ivanova

INTRODUCTION

The effectivity of the metabolic processes in the cell depends on coordination in the activity of the enzymes, the proper quantity of which must be synthesized in due time and be present in the particular compartments. This may be realized by the intracellular structure as well as by the system of gene expression, regulated by both extra- and intracellular factors. RNases, using RNA as substrate, can be involved in the regulatory processes at different stages: replication, transcription and translation of the gene material.

The present review deals with the intracellular RNases and their role in the fungal metabolism.

SUBSTRATE SPECIFICITY OF RNases

RNases may be referred to as phosphodiesterases which split the internucleotide bonds in ribonucleic acids. By the current enzyme nomenclature they belong to hydrolases E.C. 3.1.4 (Enzyme nomenclature, 1972). Table 1 shows a compilation of the different RNases. According to the mode of action RNases are divided into phosphotransferases and hydrolases (Barnard, 1969). Phosphotransferases catalyse two successive reactions: transphosphorylation with the nucleotide 2', 3'-cyclic phosphate formation, and their hydrolysis to nucleoside-3'-monophosphate (Fig. 1). RNA-hydrolases directly attack phosphodiester bonds in RNA (EC 3.1.4.20).

TABLE 1

Ribonucleases		
1.	Ribonucleate 3'- or 5'-nucleotido-hydrolase	(EC 3.1.4.20)
2.	Ribonucleate 2'-transferase (cyclisizing)	
2.1.	Non-base specific RNase	(EC 3.1.4.23)
2.2.	Pyrimidine-specific RNase	(EC 3.1.4.22)
2.3.	Purine-specific RNase	
2.4.	Guanine-specific RNase	(EC 3.1.4.8)
2.5.	Adenine-specific RNase	
3.	Special groups of RNases	
3.1.	RNA·DNA-specific RNase	(EC 3.1.4.34)
3.2.	Double-strand-specific RNase	(EC 3.1.4.24)
3.3.	Nucleotide-block-specific RNase	

There are pyrimidine-, guanine-, purine-, adenine-specific and non-base specific RNases according to their specificity for bases, adjacent to the split phosphodiester bonds. Some ribonucleases are able to recognize the secondary substrate structure and the other-nucleotide blocks. Besides, we know RNases degrading the RNA moiety of RNA·DNA hybrides.

Information about RNA-depolymerases of microbial eukaryotic cells are summed in a number of reviews (Anfinsen, White, 1961; Shapot, 1968; Barnard, 1969; Egami & Nakamura, 1969; Uchida & Egami, 1971; Bezborodov, 1974).

CHARACTERISTICS OF THE INTRACELLULAR RNases

To discuss the question about the role of RNases in the fungal metabolism it is necessary to have data on these enzymes (substrate specificity, localization) and on turnover of their substrates in the cells. The review of Egami & Nakamura (1969) shows the wide presence of RNases in the realm of fungi. At present data on fungal RNases mainly con-

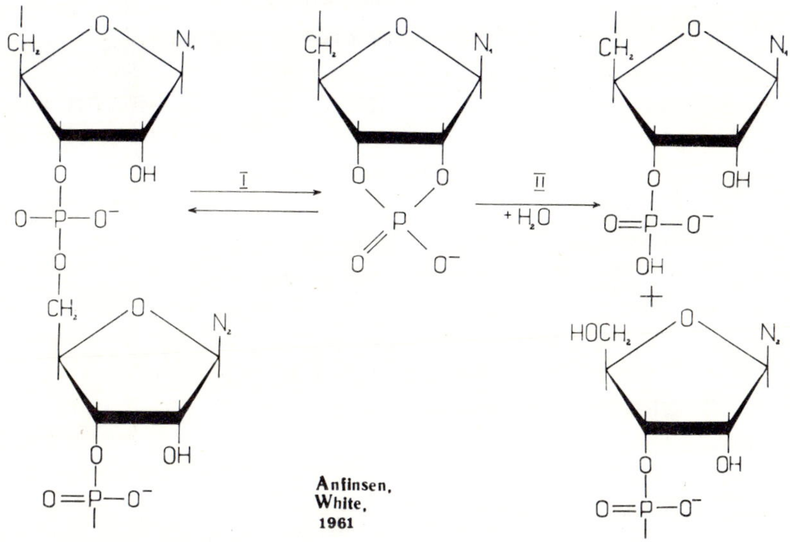

Fig. 1. Mode of action of phosphotransferases.

cern extracellular enzymes performing a digestive function. As to intracellular enzymes which are very interesting from the physiological point of view they have been studied far less. Literature so far cited was as a rule devoted to changes in the level of enzyme activity during growth and, to a lesser degree, in obtaining the purified enzymes as well as properties and specificity of the intracellular RNases.

The experimental data obtained for *Aspergillus* and *Penicillium* (Ivanova *et al.*, 1974a) show that the isoenzymic spectrum of the intracellular RNases contains all the components of the extracellular RNases and also a number of enzymes specific for the cell content. Therefore, when discussing the intracellular processes we can use data concerning the properties of extracellular enzymes. The possibility of some changes in the molecules of the enzymes during secretion must, of course, be taken into consideration. For example, extracellular *Aspergillus clavatus* RNase C_2 and its intracellular precursor RNase Ac_2 have the minor differences in both carbohydrate components and molecule conformation, the latter altering kinetic parameters of enzymic reactions (Ivanova & Valjukaite, 1976). But guanine specificity is left unchanged for both enzymes.

Table 2 represents data on the specificity of intracellular RNases of fungi and yeast. According to these data all the RNases mentioned in Table 1 were found in the fungi, and some of them such as ds-specific and block-specific RNases were inferred by analogy with prokaryotic cells (Robertson et al., 1968; Pace & Sorin, 1975) and eukaryotic cells (Rech et al., 1976; Koski et al., 1976). Pyrimidine-specific RNases have not yet been found in microorganisms. This fact seems to reflect a division of polynucleotide-splitting functions between micro- and macroorganisms having pyrimidine-specific RNases.

Table 3 shows some properties of fungal and yeast RNases. As a rule the molecular weight of the enzymes ranges from 11000 to 40000. By pH optimum the enzymes may be divided into acid and alkaline RNases.

LOCALIZATION OF INTRACELLULAR RNases

Our knowledge of RNase localization is very scanty. On one hand it may be explained by the methodical difficulties because of the low level of RNases in the cell. On the other there do not exist the specific substrates for the study of RNase localization by electron microscopy.

The method of differential centrifugation of cell homogenates with the following analysis of RNase activity is widely used. This approach was applied for studying RNase distribution in the subcellular fractions of A. niger (Vasu et al., 1971). Some enzymes with a pH optimum in the range of 6.2 - 6.8 were localized in the pellets by centrifugation at 800 g; 4000 g; 15000 g and 105, 000 g. RNases with a pH optimum of 4.0 - 4.4 predominated in the postribosomal fraction. Free and structure-bound enzymes differred by electrophoretic mobility.

The essential RNase activity of 48 h A. clavatus mycelium was revealed in 105, 000 g supernatant and only 15 per cent of the activity was confined to pellets (Ivanova et al., 1976). Guanine-specific RNase of A. clavatus was released from the periplasmic space during protoplasts for-

TABLE 2

Substrate Specificity of Intracellular RNases of Fungi and Yeast

Microorganism	Enzyme	Substrate	Action	Specificity	Products	Reference
Aspergillus clavatus	RNase Ac_1	RNA	endo	G-sp	Oligo-, 3'Gp, Gp!	Ivanova et al. (1973)
	RNase Ac_2	RNA	endo	G-sp	Oligo-, 3'Gp, Gp!	Ivanova & Valiukaite (1976)
	RNase Ac_3	RNA	endo	Non-sp	Oligo-, 3'Np, Np! A > G > U > C	Ivanova & Krupianko (1976)
Penicillium claviforme	RNase Pc_1	RNA	endo	Non-sp	Oligo-, 3'Np, Np! A > G > U > C	Ivanova et al. (1977a)
	RNase Pc_2	RNA	endo	Non-sp	Oligo-, 3'Np, Np! A > G > U > C	dto
Trichoderma koningii	RNase I	RNA	endo	Non-sp	Oligo-, 3'Np, Np! A > U > G > C	Harada & Irie (1974)
	RNase II	RNA	endo	Non-sp	Oligo-, 3'Np, Np! A > U > G > C	dto
	RNase III	RNA	endo	Non-sp	Oligo-, 3'Np, Np! G > A > U > C	dto
Achlya bisexualis	RNase(-s)	RNA polyA·mRNA	Not known			Sutherland et al. (1976)
Neurospora crassa	RNase N_2	RNA	endo	Non-sp	Oligo-, 3'Np, Np!	Takai et al. (1967)
	RNase N_3	RNA	endo	G-sp	Oligo-, 3'Np, Gp!	dto
Physarum polycephalum	RNase P_{p-1}	RNA	endo	G-sp	Oligo-, 3'Gp, Gp!	Hiramaru et al. (1969a)
	RNase P_{p-2}	RNA	endo	Non-sp	Oligo-, 3'Np, Np!	dto
	RNase P_{p-3}	RNA	endo	Non-sp	Oligo-, 3'Np, Np!	dto
	RNase P_{p-4}	RNA	endo	Non-sp	Oligo-, 5'Np A, G > C, U	Hiramaru et al. (1969b)
Candida lipolytica	RNase Cl	RNA	exo	Non-sp	Np!	Imada et al. (1975)
Saccharomyces cerevisiae	RNase	RNA	endo	Non-sp	Oligo-, Np!	Ohtaka et al. (1963)
	RNase H_1	RNA·DNA	exo	Non-sp	5'Np	Wyers et al. (1976a, b)
	RNase H_2	RNA·DNA	endo	Non-sp	Oligo-, 5'Np	dto

TABLE 3

Properties of Intracellular RNases of Fungi and Yeast

Microorganism	Enzyme	pH optimum	Mol. weight	Localization	Reference
Aspergillus clavatus	RNase Ac_1	8.0	11500	Periplasmic space	Ivanova & Valiukaite (1974b)
	RNase Ac_2	8.0	13000	Ribosomes; Periplasmic space	Ivanova et al. (1976)
	RNase Ac_3	4.7	26000	Lyzosomes	Ivanova et al. (1977b)
Penicillium claviforme	RNase Pc_1	6.5	38000	-	Ivanova et al. (1977a)
	RNase Pc_2	5.8	36000	-	dto
Trichoderma koningii	RNase I	4.3	24000	-	Harada & Irie (1974)
	RNase II	4.3	18000	-	dto
	RNase III	4.8	6000	-	dto
Achlya bisexualis	RNase (-s)	5.0 - 6.0	-	Lyzosomes	Sutherland et al. (1976)
Neurospora crassa	RNase N_2	8.0	36000	-	Takai et al. (1967)
	RNase N_3	6.0 - 7.0	10000	-	dto
Neurospora crassa	RNase	5.0 - 5.5	-	Ribosomes	Ursino et al. (1969)
	RNase	6.0 - 8.0	-	Ribosomes	dto
Aspergillus oryzae	RNase	6.0	16000	-	Horikoshi (1971)
Physarum polycephalum	RNase P_{p-1}	6.7	40000	-	Hiramaru et al. (1969a, b)
	RNase P_{p-2}	4.5	40000	-	dto
	RNase P_{p-3}	5.5	11000	-	dto
	RNase P_{p-4}	4.0	-	-	dto
Candida lipolytica	RNase Cl	5.5 - 6.0	25000	-	Imada et al. (1975)
Lentinus edodes	RNase I	4.5	-	-	Mauri et al. (1972)
	RNase II	4.5	-	-	dto
	RNase IV	7.5 - 8.0	-	-	dto
Saccharomyces cerevisiae	RNase H_1	5.5 - 8.0	21000	-	Wyers et al. (1976)
	RNase H_2	8.0 - 8.5	48000	-	dto

mation. The findings that guanine-specific RNase is mainly located in the periplasmic space, that it is present at all growth stages, that it is very closely related to extracellular RNase C_2 by its physical and chemical properties testify that intracellular RNase Ac_2 is the precursor of extracellular RNase C_2 (Ivanova *et al.*, 1976). Non-base specific RNase Ac_3 of *A. clavatus* was found in the subcellular fraction, which may be considered as lysosomes by its enzymic composition and electronmicroscopic characteristics (Ivanova *et al.*, 1977b).

Preliminary data were obtained about the RNase activity in the nuclear, microsomal and lysosomal fractions of the homogenates of *Achlya bisexualis* (Sutherland *et al.*, 1976). The presence of two RNases with pH optima of 5.5 and 6.5 in the ribosomes of *Neurospora crassa* (Ursino *et al.*, 1969), was shown. Ribosomal localization of RNases was revealed in the mycelium of *A. niger* (Vasu *et al.*, 1971) and *A. clavatus* (Ivanova *et al.*, 1976). In the mitochondrial fractions of *Neurospora crassa* there are two forms of RNA-depolymerases, one of which is bound to membranes and the other exists in a free form (Martin & Wagner, 1975).

Thus, the data on localization of the intracellular RNases permit us to conclude that structure-bound RNases are present in fungi. Such compartmentation directs RNase activity towards definite substrates.

RNases IN THE RNA METABOLISM

Though our knowledge about RNases of fungi is rather fragmentary we can put forward some suppositions about the RNase involvment in the metabolism of nucleic acids knowing RNase specificity and their intracellular localization. Multiple molecular forms of intracellular RNases in fungi (Vasu *et al.*, 1972; Ivanova *et al.*, 1974a; Chet *et al.*, 1973) point to a variety of functions in the cells.

In the process of DNA replication RNA·DNA-specific RNases (Banks, 1974; Wyers *et al.*, 1976b) may take part in the initiation of DNA synthesis by splitting off RNA segments (Sugino *et al.*, 1972; Berkover, 1973). Nuclear localization shown for these enzymes testify to this supposition (Tashiro et al., 1976).

For bacterial RNA·DNA-specific RNases another possible function is discussed, namely, participating in conformation changes in chromatin structure during the cell cycle (Wyers et al., 1976a). RNA molecules were found in the bacterial genome stabilizing DNA folds (Drlica & Worcel, 1975). As for eukaryotic cells this function of RNases was not yet confirmed.

Transcription is the next link in transferring information from genes. The enzymes mentioned above, may participate in regulation of RNA biosynthesis at this stage. There is evidence for a stimulating action of RNase on transcription carried out by RNA-polymerase (Natori et al., 1973; Wyers et al., 1976a). In connection with this it is very interesting to note that in the yeast cells RNase H activity, associated with RNA-polymerase A, have been found (Huet et al., 1976). The possible enzyme function consists in splitting DNA·RNA hybrids which may be formed at the sites of initiation of DNA transcription.

It was found that transcribed RNAs from genes were longer than translated mRNA and than tRNA and rRNA working in the ribosomes. The processing of the precursors into mature RNAs includes nucleoside modification, the shortening of nucleotide chains and the addition of the specific nucleotide sequences at 3'- and 5'-terminal molecules. The detailed discussion of nucleic acid-processing in prokaryotes is given in many reviews (Altman & Robertson, 1973; Levy, 1975; Datta & Niyogi, 1976).

Knowledge of the structure of pre-tRNA in prokaryotes allows the sites of cleavage in pre-tRNA to be identified. At least two enzymes, RNase P and RNase Q, are involved in mature tRNA formation (Robertson et al., 1972; Sakano & Shimura, 1975). The stepped character of processing precursor molecules supposes recognition by enzymes rather than by molecule configuration of definite nucleosides close to the split phosphodiester bonds (Perry, 1976).

It is known that the tRNA biosynthesis in eukaryotic cells also passes through precursor formation (Bernhard & Darnell, 1969). From the experiments with pre-tRNA of yeast it is obvious that there exists the intracellular specific enzyme activity yielding tRNA from the precursors by endonucleolitic splitting (Blatt & Feldman, 1973). For lack of data on pre-tRNA processing in fungi we are obliged to turn to the scanty examples among other eukaryotic organisms. Pre-tRNA of the silk glands of Bombyx mori is converted to tRNA by the enzyme of E. coli (Chen & Siddigui, 1973). The same splitting in precursor molecules was observed in the presence of the ribosomal crude extract of the silk glands of Bombyx mori (Tsutsumi et al., 1974). Experiments on splitting pre-$tRNA^{tyr}$ of E. coli by the enzyme from KB cells indicated the identity of specificity of this enzyme and RNase P of E. coli. Availability of the enzyme in the cytoplasm shows the possible localization tRNA "maturation" in the cytoplasm of eukaryotic cells (Koski et al., 1976).

These data testify to the presence of the enzymes involved in tRNA processing in eukaryotic cells. The striking similarity in enzyme specificity of pro- and eukaryotic cells points to the fact that their pre-tRNA conformations are very much alike. According to Perry (1976) substrate conformation defines enzyme-substrate recognition.

The rRNA synthesis in fungi and yeast runs in the same way as in higher eukaryotic cells. The initial transcript 36S-38S rRNA contains the sequences 17S-18S rRNA, 25S-28S rRNA and 5.8S RNA (Trapman & Planta, 1976; Pearson & Thomas, 1976). Sequences of rRNA location in the precursor molecules of Neurospora crassa and Saccharomyces carlsbergensis are as follows: 5'-17S rRNA - 25S rRNA - 3' (Trapman & Planta, 1975; Russell et al., 1976). Molecule-precursors undergo a number of changes, shown schematically in Fig. 2 (Trapman et al., 1976; Trapman & Planta, 1976). At all stages of processing rRNA associates with proteins forming the nucleoprotein complexes. 5S RNA seems to be transcribed not depending on 37S rRNA and linked to 66S particles.

It is clear that rRNA processing passes through several stages. The staged sequence as in the case with tRNA is explained by the fact that the action of one enzyme on pre-rRNA creats conditions for recognition and catalysis by another enzyme (Perry, 1976). In spite of many pretenders to pre-rRNA splitting there is no direct evidence for definite

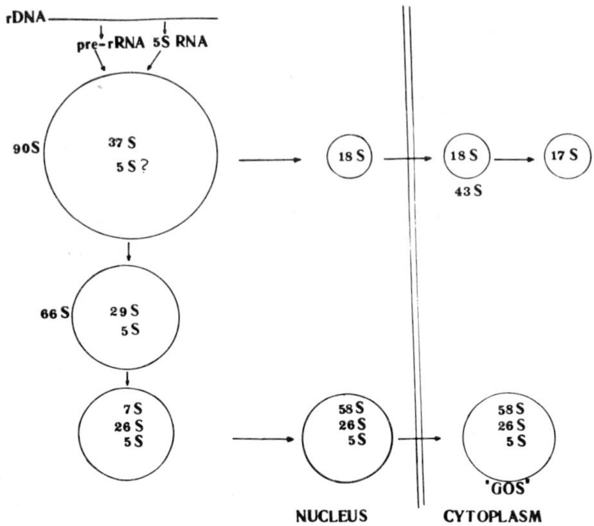

Fig. 2. Scheme of ribosome maturation in yeast (Trapman *et al.*, 1976).

participants in this process both for eukaryotes and prokaryotes. At some initial stages the involvment of the enzymes specific to double-stranded structures (enzymes similar to RNase III (Schweitz & Ebel, 1971) of *E. coli* is very probable.

In the cells of the higher eukaryotes the activity is determined for the enzymes having a strong specificity to ds structure (Robertson & Matthews, 1973; Rech *et al.*, 1976) and for enzymes with marked preferences to the secondary structure of RNA (Palese & Koch, 1972; Libonati *et al.*, 1975). If we consider data on the influence of high density of the negative charges on enzyme molecules upon their ability to split the secondary structure of RNA then RNases with an isoelectric point at pH 8.0 - 9.0 turn out to possess such an activity (Libonati *et al.*, 1976). The presence of RNA-depolymerase with an isoelectric point below pH 8.0 in the mycelium of *P. claviforme* was revealed by isoelectric focusing (Ivanova, 1977). Besides, the property to split ds structure of RNA occurs due to dimeric forms of RNases

(Libonati et al., 1975). When ionic strength is low, some of the RNases are able to aggregate. It can take place in vivo.

Localization of the site of splitting by RNases in ribonucleoprotein complexes of pre-rRNA is provided seemingly by protein shielding of nucleotide sequences. Otherwise, protein absence results in nonspecific splitting along the whole polynucleotide chain (Winicov & Perry, 1974; Hayes & Vasseur, 1977).

Ribosomal RNases of Neurospora crassa (Somberg & Davis, 1965) are inactive towards rRNA in the intact ribosomes, but they readily split rRNA when dissociation of the ribosomal subunits occurs. Therefore, their participation in the catabolic splitting of RNA is more probable. Specificity of RNase Ac_1 and Ac_2 of A. clavatus is more pronounced for rRNA. Only 20 % of acid soluble products from rRNA in comparison to 55 % from tRNA are formed in vitro during 24 h incubation (Ivanova et al., 1974c).

Heterogeneous nuclear RNAs (hn RNA) are precursors of mRNAs in eukaryotic cells. A number of reviews are devoted to their metabolism (Weinberg, 1973; Darnell et al., 1973; Perry et al., 1975). Unlike the prokaryotes pre-mRNA of the eukaryotes undergo some alterations in nucleus and after that they enter the cytoplasm. It is supposed that polynucleotide sequences of mRNA are located near 3'-termini of pre-mRNA, and 5'-termini sequences are destroyed when mRNA is formed (Georgiev, 1969; Georgiev et al., 1972). As for RNases involved in pre-mRNA processing we know nothing. There are but suggestions about incorporation of enzymes, splitting ds structures of RNA. In this process hairpin structures in pre-mRNA may be the places of recognition for RNases (Ryskov et al., 1976).

Adenylation of pre-mRNA at 3'-OH termini is one of the stages of mRNA maturation. Biosynthesis of these sequences is performed enzymatically (Rose & Jacob, 1976; Antoniades & Antonoglou, 1976). Nuclear poly A(+) RNAs of Saccharomyces cerevisiae have molecular weight $1.3 \cdot 10^6$ daltons which is twice that of cytoplasmic poly A(+) RNA (Groner & Phillips, 1975). Hence it is necessary to have additional splittings occurring in pre-mRNA molecules before their leaving the nuclei. The length of the polyadenyl sequences varies. For example, the poly A segments in zoospores of Blastocladiella emersonii are on the average 100 nucleotides (Schimmelpfeng et al., 1976), of Trichoderma 50 - 60 nucleotides (Rosen & Edelman, 1976), of Saccharomyces cerevisiae 20 - 60 nucleotides (Groner & Phillips, 1975). Many studies show the protective role of poly A segments for maintenance of biological activity of mRNA molecules (Spector & Baltimore, 1974). The effectiveness of inhibition of RNase activity decreases with shortening poly A chains (Hieter et al., 1976).

Fig. 3 gives the scheme of mRNA synthesis (Müller et al., 1976). Poly A(+) RNAs penetrate only partly into the cytoplasm, the major quantity being degraded in the nuclei. According to the scheme poly A specific endonuclease splits poly A(+) RNA (Müller, 1976) and then RNA sequences are removed non-specifically. In this way adenyl specific endoribonuclease is involved in the control of poly A synthesis in cell nuclei as well as in the control of mRNA transfer into the cytoplasm.

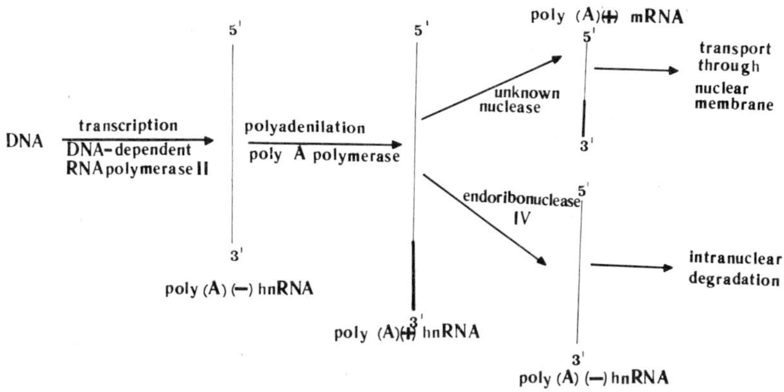

Fig. 3. Tentative model describing the enzymic steps of non-histone mRNA synthesis (Müller et al., 1976).

While considering RNA synthesis during the cell cycle of Saccharomyces cerevisiae a correlation turned out to exist between gene number and rate of RNA and poly A(+)

RNA transcription (Fraser & Carter, 1976). The rate of RNA and poly A(+) RNA synthesis is constant in the G_1 phase and doubles in the S phase, after that being constant at that new level. Moreover, the rate increase in poly A(+) RNA corresponds to the acceleration of its precursor processing.

The presence of $m^7G(5')pppXp$ at 5'-termini of polynucleotide chains are characteristic of all eukaryotic mRNAs. There are suggestions about a role of these sequences in initiation of protein synthesis (Both _et al._, 1975). As for yeast, A or G may occupy the X position (Sripati _et al._, 1976; De Kloet & Andrean, 1976) and for higher eukaryotes more often 2'-O-methylated A, U, G, C are the nucleosides formed (Adams & Cory, 1975). These dinucleotide structures are not subjected to splitting by RNase because 5'-5' phosphodiester bonds are resistant to RNase action (Ivanova _et al._, 1974d, e). But there are many other enzymes such as polynucleotide kinases which remove $m^7G(5'p)$ sequences (Abraham & Lillehaug, 1976).

CHANGES IN THE ISOENZYMATIC RNase SPECTRA AND IN THE ACTIVITY DURING GROWTH

The idea about heterogeneous systems of RNA-depolymerases appeared thanks to electrophoretic, chromatographic and sedimentation methods of investigation. According to modern concepts multipie forms of enzymes are considered as genetically determined enzyme forms or as conformational modification of the same molecules.

The ability of many fungi to produce simultaneously several extracellular RNases differing from each other by physico-chemical properties was marked. For example: RNase T_1 and T_2 of _A. oryzae_ (Egami _et al._, 1971); RNase U_1, U_2, U_3, U_4 of _Ustilago sphaerogena_ (Glitz & Dekker, 1964; Arima _et al._, 1968); four non-specific and one guanine specific RNases of _P. brevicompactum_ (Bezborodova & Bagdasarjan, 1972); RNases R_1 and R_2 of _Rhizopus oligosporus_ (Woodroof & Glitz, 1971) RNases F1 and F_2 of _Fusarium moniliforme_ (Omori _et al._, 1972).

Heterogeneity of the composition of the intracellular RNases is also observed. The presence of 12 RNases in the cellular homogenate of *Physarum polycephalum* was revealed by an isoelectric focusing method (Chet *et al.*, 1973). Vasu in his works showed that fungi belonging to the genus *Aspergillus* produced 5 - 7 intra- and 2 - 3 extracellular RNA-depolymerases. RNase enzymograms may serve as taxonomic criteria in the genus *Aspergillus* (Vasu *et al.*, 1972).

The isoenzymic spectrum of RNases changes during submerged cultivation of fungi. The whole set of enzymes produced by the culture does not depend on changes in the medium. Under these conditions it is activity and time of appearance of enzymes that alters, as was shown for *P. claviforme* (Ivanova *et al.*, 1975). It should be noted that RNase activity is present during the whole time of cultivation and that heterogeneity of RNases is more pronounced during the initial 1 - 2 days growth of *P. claviforme*. This period corresponds to an increase in intracellular RNase activity which probably may convert nucleic acids and their precursors during the intense intracellular biosynthesis (Ivanova *et al.*, 1975).

Sporulation, resting spores and their germination are very convenient objects and means for studying differentiation. During Ca-induced sporangium formation in *Achlya bisexualis* the biosynthesis of nucleic acids and proteins is required. The rate of protein synthesis increases twice compared to that of synthesis in the vegetative mycelia and is maintained high during differentiation (Timberlake *et al.*, 1973). Biochemical and morphological changes are accompanied by those of RNA-depolymerase activity. Decrease in RNase activity is completed prior to cross wall formation and then the RNase activity increases again with subsequent reduction at spore discharge. A correlation exists between decrease in RNase activity and increase in cellular poly A (+) RNA (Sutherland *et al.*, 1976). As was shown the RNase activity depends on transcription and translation.

There are data on relation between RNase activity and turnover of nucleic acids during sporulation of *Saccharomyces cerevisiae* (Tsuboi, 1976; Betz & Weiser, 1976). Increase in RNase activity and breakdown of vegetative RNA are the early events during sporulation (Tsuboi & Yanagishima, 1976). Sporulation in *Saccharomyces cerevisiae* involved meiosis. Degradation products of vegetative RNA are used for premeiotic DNA synthesis (Tsuboi, 1976). There is evidence that degradation of vegetative RNA is not a main source for RNA synthesis during sporulation. For this purpose internal reserves may be used (Chaffin et al., 1974).

Distinct phases were observed in producing and developing sclerotia of rice sheath blight (Hashiba & Staples, 1976). Reorientation of the cellular metabolism was accompanied by a rapid RNA synthesis and decline in the contents of poly A(+) RNA in the first phase. RNase activity, proteinase activity and ribosome number increased in the second phase. Apparently, activation of the intracellular reserves by hydrolytic enzymes occurred owing to a sharp decrease in permeability of sclerotia.

Experiments have shown with certainty that all the participants of protein synthesis including nucleic acids are present in the fungal spores. Some reviews are devoted to these subjects (Sussman & Douthit, 1973; Van Etten et al., 1976). Sometimes RNA biosynthesis begins immediately after putting spores into the growth medium (Roheim et al., 1974), and sometimes biosynthesis shifts from 10 to 240 min (Van Etten et al., 1976). There is the opinion that biosynthesis of the early proteins during spore germination is performed at the expense of mRNA present in the resting spores (Hollomon, 1973). The sequences of mRNA, tRNA, rRNA synthesis during spore germination varies with different fungi. For example, *A. oryzae* has the following sequence: rRNA, tRNA, mRNA (Ono et al., 1966); *Botryodiplodia theobromae*: mRNA, tRNA, rRNA (Knight & Van Etten, 1976); *Peronospora tabacina*: rRNA, mRNA, tRNA (Hollomon, 1970).

Ribonucleic acids are synthesized as precursors followed by processing to "mature" RNA. So pre-rRNA ($2.25 \cdot 10^6$ daltons) and two rRNAs ($1.4 \cdot 10^6$ and $0.7 \cdot 10^6$ daltons) are found in germinating spores of *Phycomyces blakesleeanus* (Van Assche & Carlier, 1973). The rate of mRNA splitting influences the spectrum of the protein synthesized. Change in the quantity of A(+) RNA during spore germination of *Botryodiplodia theobromae* are shown (Knight & Van Etten, 1976a, b). 45 - 50 % of the whole population of mRNA do not contain poly A sequences. As was mentioned above poly A(+)

and poly A(-) RNA have different half-lives.

The data presented indicate that RNases must have been present both in the resting spores and during their germination. Isoenzymic spectra of RNA-depolymerases of the resting spores and vegetative mycelium of _P. claviforme_ are the same with the exception of one additional component in the spores. But RNA-depolymerases which were the main components in the isoenzymic spectrum of vegetative mycelium became the minor components in conidiospores. Occurrence of only some of the RNases from those present in mycelium was found in conidia of _A. clavatus_. Two guanine specific RNases and one non-specific RNase which is localized in lysosomes at the stage of vegetative mycelium are present both in conidia and in mycelium. Moreover, there is RNase with high electrophoretic mobility in conidia. The major proportion of the ribonuclease activity falls into this class of RNases. The enzyme is absent in mycelium at submerged cultivation during 3 days (Ivanova, 1977).

Thermolabile and latent RNases, the latter is bound to a protein inhibitor, are found in conidia of _A. oryzae_. The activity of latent enzyme disappeares during spore germination. Role and specificity of this RNase is not known. The enzyme was not revealed in vegetative mycelium (Horikoshi, 1971).

Thus, there are RNases specific for resting spores. It is likely that it reflects the unknown features of nucleic acid turnover in spores.

FACTORS LIMITING RNase ACTIVITY

The intracellular RNase activity is controlled, first, by the regulation of the RNase biosynthesis and their secretion from the cells, and second, by the inhibition or activation of the enzymes already existing. The questions concerning the regulation of RNase activity are discussed in some reviews (Egami & Nakamura, 1969; Ivanova, 1974g). In literature there are data on the influence of medium components on RNase activity (Glitz & Dekker, 1964; Arima _et al._, 1968; Fujishima _et al._, 1972; Ivanova _et al._, 1974f; Hasunuma _et al._, 1976). It is shown that extracellular RNases of some fungi belong to the inducible enzymes (Yanagida et al., 1964; Arima et al., 1968; Glitz et al., 1972).

I should like to dwell upon the factors affecting RNase activity already existing in the cell (Table 4). Activity regulation of RNase may occur by conversion of substrates (RNA) to a form not split by RNases; by alteration in the active centre of the enzyme itself; or by compartmentation of enzymes and substrates in the cell.

TABLE 4

Factors, Limiting RNase Activity

1. Compartmentation of RNases and their substrates
2. Protein-substrate association
3. Substrate conformation
4. Substrate modification
   a) polyadenylation
   b) methylation
5. Natural inhibitors
   a) high molecular
   b) low molecular

The intracellular structural elements associated with enzymes play a significant role in the regulation of activity and specificity of RNases. As we mentioned earlier ribosomes of *N. crassa* protect rRNA against ribosomal RNase action (Somberg & Davis, 1965); pre-rRNA binding with ribosomal proteins determines activity. Pre-rRNA transformation into rRNA slows down by limiting the synthesis of ribosomal proteins of *Saccharomyces cerevisiae* (Wejksnora & Haber, 1976). I have already mentioned the influence of the ribonucleoprotein complex formation on the location of activity of RNase in polynucleotide sequences of precursors.

A part of my report was devoted to the cell compartmentation of RNases which is very essential in restricting RNase activity.

The secondary structury of the nucleic acids is inaccessible to some RNases such as: T_1 of *A. oryzae*, *A.*

fumigatus RNase (Glitz *et al*., 1974).

The fine substrate specificity of intracellular RNases of *A. clavatus* was investigated with the help of nucleoside cyclophosphates modified in the base and in the ribose and with the help of dinucleoside phosphate analogues. It appeared that many nucleotide modifications which exist in the natural rRNA and tRNA were critical for RNase action (Ivanova *et al*., 1974d, e). The protective role of the polyadenyl sequences at 3'-termini of mRNA have been discussed above.

At present it is known that the fungi have natural protein inhibitors some of which being highly specific (Ivanova & Bezborodov, 1973) and the other non-specific, inhibiting pancreatic RNase A (Argyrakis & Vomvoyanni, 1973). There are evidences for a direct inhibition of RNase activity by nucleotides as well as for an indirect action of nucleotides on RNases through activition of the protein inhibitor of RNases as was shown for RNase of *Candida lipolytica* (Imada *et al*., 1972a, b).

Polyamines (putrescine, spermidine, spermine) found in the different eukaryotic cells including fungi (Wither & Stevens, 1976) are factors activating RNases. The mechanism of the polyamine action on the hydrolytic activity is obscure: polyamines may interact with RNA or may influence the conformation of the active centre (Yanagawa *et al*., 1966). Spermine and spermidine ability causing dissociation of RNase-polynucleotide complexes suggest their participation in the regulation of nucleic acid degradation (Frank *et al*., 1976).

Summing up I should like to note that RNases are still the white spots in the map of the nucleic acid metabolism. Even in the case of the obvious correlation between RNase activity and nucleic acids we cannot pinpoint the exact localization of RNase action. Approaches, successfully used for studying the RNase role in the nucleic acid metabolism of bacteria, may be applied also to fungi. Individual enzymes, mutants with definite RNases and substrates with nucleotide sequences already mentioned are likely to bring us nearer to understanding RNase function in cellular differentiation.

REFERENCES

Abraham, K. A. & Lillehaug, J. R. (1976). FEBS Lett. 71, 49.

Adams, J. M. & Cory, S. (1975). Nature 255, 28.

Altman, S. & Robertson, H. D. (1973). Mol. Cell. Biochem. 1, 83.

Anfinsen, C. B. & White, T. H. (1961). In: The Enzymes. Boyer, P. D. (ed.) Academic Press, New York, 95.

Antoniades, D. & Antonoglou, O. (1976). Biochim. et Biophys. Acta 474, 609.

Argyrakis, M. & Vomvoyanni, V. (1973). Biochim. et Biophys. Acta 299, 300.

Arima, T., Uchida, T. & Egami, F. (1968). Biochem. J. 106, 609.

Banks, G. R. (1974). Eur. J. Biochem. 47, 499.

Barnard, E. A. (1969). Ann. Rev. Biochem. 38, 679.

Berkover, I., Leis, J. & Hurwitz, J. (1973). J. Biol. Chem. 248, 5914.

Bernhard, D. & Darnell, J. E. (1969). J. Mol. Biol. 42, 43.

Betz, H. & Weiser, U. (1976). Eur. J. Biochem. 70, 385.

Bezborodov, A. M. (ed.) (1974). "Nukleazy Mikroorganizmov" Nauka, Moscow.

Bezborodova, S. I. & Bagdasarjan, Z. N. (1972). Mikrobiologya 41, 773.

Blatt, B. & Feldman, H. (1973). FEBS Lett. 37, 129.

Both, G. W., Banerjee, A. K. & Shatkin, A. J. (1975). Proc. Natl. Acad. Sci. USA 72, 1180.

Chaffin, W. L., Sogin, S. J. & Halvorson, H. O. (1974). J. Bacteriol. 120, 872.

Chen, G. S. & Siddigui, M. A. Q. (1973). Proc. Natl. Acad. Sci. USA 70, 2610.

Chet, I., Retig, N. & Henis, Y. (1973). Biochim. et Biophys. Acta 294, 343.

Darnell, J. E., Jelinek, W. R. & Molloy, G. R. (1973). Science 181, 1215.

Datta, A. K. & Niyogi, S. K. (1976). In: Progress in Nucleic Acid Research and Molecular Biology, v. 17, 271.

De Kloet, S. R. & Andrean, B. A. G. (1976). Biochim. et

Biophys. Acta 425, 401.
Drlica, K. & Worcel, A. (1975). J. Mol. Biol. 98, 393.
Egami, F. & Nakamura, K. (1969). Microbial Ribonucleases. Springer-Verlag, Berlin, New York.
Enzyme nomenclature, Recommendations (1972) of the International Union of Pure and Applied Chemistry and the International Union of Biochemistry. Elsevier, Netherlands (1973).
Frank, J. J., Hawk, I. A. & Levy, C. C. (1976). Biochim. et Biophys. Acta 432, 369.
Fraser, R. S. S. & Carter, B. L. A. (1976). J. Mol. Biol. 104, 223.
Fujishima, T., Suzuki, M. & Yoshino, H. (1972). J. Ferment. Technol. 50, 465.
Ghosh, R. K. & Thangamani, A. (1973). Can. J. Microbiol. 19, 1219.
Glitz, D. G. & Dekker, C. A. (1964). Biochemistry 3, 1391.
Glitz, D. G., Eichler, D. C. & Angel, L. (1974). Anal. Biochem. 62, 552.
Glitz, D. G., Angel, L. & Eichler, D. C. (1972). Biochemistry 11, 1746.
Georgiev, G. P. (1969). J. Theoret. Biol. 25, 473.
Georgiev, G. P., Ryskov, A. P., Coutelle, Ch., Mantieva, V. L. & Avakyan, E. R. (1972). Biochim. et Biophys. Acta 259, 759.
Groner, B. & Phillips, S. L. (1975). J. Biol. Chem. 250, 5640.
Harada, M. & Irie, M. (1974). Chem. Phyrm. Bull. 22, 1850.
Hashiba, T. & Staples, R. C. (1976). J. Gen. Microbiol. 96, 239.
Hasunuma, K., Toh-E. A. & Ishikawa, T. (1976). Biochim. et Biophys. Acta 432, 223.
Hayes, F. & Vasseur, M. (1976). Eur. J. Biochem. 61, 433.
Hieter, P. A., Lebendre, S. M. & Levy, C. C. (1976). J. Biol. Chem. 231, 3287.
Hiramaru, M., Uchida, T. & Egami, F. (1969a). J. Biochem. 65, 693.
Hiramaru, M., Uchida, T. & Egami, F. (1969b). J. Biochem. 65, 701.
Hollomon, D. W. (1970). J. Gen. Microbiol. 62, 75.

Hollomon, D. W. (1973). J. Gen. Microbiol. 78, 1.
Horikoshi, K. (1971). Biochim. et Biophys. Acta 240, 532.
Huet, J., Wyers, F., Buhler, J.-M., Sentenac, A. & Fromageot, P. (1976). Nature 261, 431.
Imada, A., Sinskey, A. J. & Tannenbaum, S. R. (1972a). Proc. IV IFS Ferment. Technol. Today (Terui, G., ed.) Society of Fermentation Technology, Japan 455.
Imada, A., Sinskey, A. J. & Tannenbaum, S. R. (1972b). Biochim. et Biophys. Acta 268, 674.
Imada, A., Hunt, J. W., Van De Sande, H., Sinskey, A. J. & Tannenbaum, S. R. (1975). Biochim. et Biophys. Acta 395, 490.
Ivanova, G. S. & Bezborodov, A. M. (1973). Folia Microbiol. 18, 212.
Ivanova, G. S., Valiukaite, R. V. & Maximova, N. K. (1973). In: Sravnitelnaja Fiziologija i Biochimija Mikroorganizmov. Gos. Univ. im. N. I. Lobachevskogo, Gorki 70.
Ivanova, G. S., Valiukaite, R. V., Youngol, L. M. & Bezborodov, A. M. (1974a). In: Proc. 1st Symp. "Ribosomes and RNA metabolism" Publishing House of the Slovak Acad. Sci., Czechoslovakia 485.
Ivanova, G. S. & Valiukaite, R. V. (1974b). Mikrobiologija 43, 417.
Ivanova, G. S., Valiukaite, R. V., Maximova, N. K. & Bezborodov, A. M. (1974c). Biochimija 39, 1234.
Ivanova, G. S., Holy, A. & Bezborodova, S. I. (1974d). Collection Czechoslov. Chem. Commun. 39, 993.
Ivanova, G. S., Holy, A., Zelinkova, E., Bezborodova, S. I., Abrosimova-Amelyanchik, N. M. & Tatarskaya, R. I. (1974e). Collection Czechoslov. Chem. Commun. 39, 2986.
Ivanova, G. S. (1974f). In: Nukleazy Mikroorganizmov (Bezborodov, A. M., ed.) Nauka, Moscow 282.
Ivanova, G. S., Youngol, L. M. & Bezborodov, A. M. (1975). In: Proc. 1st Internat. Congress of IAMS (Hasegawa, T., ed.) 5, 243.
Ivanova, G. S. & Krupianko, V. I. (1976). In: Proc. 2nd Symp. "Ribosomes and RNA metabolism" Publishing House of the Slovak Acad. Sci., Czechoslovakia, 205.

Ivanova, G. S. & Valiukaite, R. V. (1976). In: Proc. 2nd Symp. "Ribosomes and RNA metabolism" Publishing House of the Slovak Acad. Sci., Czechoslovakia, 197.

Ivanova, G. S., Youngol, L. M. & Bezborodov, A. M. (1977a). Prikl. Biochim. Microbiol. in press.

Ivanova, G. S., Markelova, N. Yu., Petrikevich, S. B. & Bezborodov, A. M. (1977b). Mikrobiologija, in press.

Knight, R. H. & Van Etten, J. L. (1976a). J. Gen. Microbiol. 95, 257.

Knight, R. H. & Van Etten, J. L. (1976b). Arch. Mikrobiol. 109, 45.

Koski, R. A., Bothwell, A. L. & Altman, S. (1976). Cell 9, 101.

Levy, C. C. (1975). Life Sciences 17, 311.

Libonati, M., Malorni, M. C., Parente, A. & D'Alessio, G. (1975). Biochim. et Biophys. Acta 402, 83.

Libonati, M., Sorrentino, S., Galli, R., La Montagna, R. & Di Donato, A. (1975). Biochim. et Biophys. Acta 407, 292.

Libonati, M., Furia, A. & Beintema, J. J. (1976). Eur. J. Biochem. 69, 445.

Martin, C. E. & Wagner, R. P. (1975). Can. J. Biochem. 53, 823.

Mouri, T. & Hashida, W. (1972). Abstr. IV Internat. Ferment. Symp. Kyoto, Japan 235.

Müller, W. E. C. (1976). Eur. J. Biochem. 70, 241.

Müller, W. E. C., Geibert, G., Steffen, R. & Lahn, R. K. (1976). Eur. J. Biochem. 70, 249.

Natori, S., Takeuchi, K., Takahashi, K. & Mizuno, D. (1973). J. Biochem. 73, 879.

Ohtaka, Y., Uchida, K. & Sakai, T. (1963). J. Biochem. 54, 322.

Omori, A., Sato, S. & Tamiya, N. (1972). Biochim. et Biophys. Acta 268, 125.

Ono, T., Kimura, K. & Kanagita, T. (1966). J. Gen. Appl. Microbiol. 12, 13.

Pace, N. R. & Sogin, M. L. (1975). Brookhaven Symp. Biol. 26, 224.

Palese, P. & Koch, G. (1972). Proc. Natl. Acad. Sci. USA 69, 698.

Pearson, J. A. & Thomas, K. (1976). Biochim. et Biophys. Acta 447, 28.
Perry, R. P. (1976). Ann. Rev. Biochem. 45, 605.
Perry, R. P., Bard, S., Hames, B. D. & Kelley, D. E. (1975). Proc. 9th FEBS Meet. Akademiai Kiado, Budapest 17.
Rech, J., Cathala, G. & Jeanteur, Ph. (1976). Nucl. Acid. Res. 3, 2055.
Robertson, H. D., Webster, R. E. & Zinder, N. D. (1968). J. Biol. Chem. 243, 82.
Robertson, H. D., Altman, S. & Smith, J. D. (1972). J. Biol. Chem. 247, 5243.
Robertson, H. D. & Matthews, M. W. (1973). Proc. Natl. Acad. Sci. USA 70, 225.
Roheim, J. R., Knight, R. H. & Van Etten, J. L. (1974). Dev. Biol. 41, 137.
Rose, K. M. & Jacob, T. (1976). Eur. J. Biochem. 67, 11.
Rosen, D. & Edelman, M. (1976). Eur. J. Biochem. 63, 525.
Russel, P. J., Hammett, J. R. & Selker, E. U. (1976). J. Bacteriol. 127, 785.
Ryskov, A. P., Kramerov, D. A. & Georgiev, G. P. (1976). Biochim. et Biophys. Acta 447, 214.
Sakano, H. & Shimura, Y. (1975). Proc. Natl. Acad. Sci. USA 72, 3369.
Schimmelpfeng, L., Grierson, D. & Zetsche, K. (1976). Biochim. et Biophys. Acta 435, 340.
Schweitz, H. & Ebel, J. P. (1971). Biochimie 53, 585.
Shapot, V. S. (1968). Nukleazy, Medicina, Moscow.
Somberg, E. W. & Davis, F. F. (1965). Biochim. et Biophys. Acta 108, 137.
Spector, D. H. & Baltimore, D. (1974). Proc. Natl. Acad. Sci. USA 71, 2983.
Sripati, C. E., Groner, Y. & Warner, J. R. (1976). J. Biol. Chem. 251, 2898.
Sugino, A., Hirose, S. & Okazaki, R. (1972). Proc. Natl. Acad. Sci. USA 69, 1863.
Sussman, A. S. & Douthit, H. A. (1973). Ann. Rev. Plant Physiol. 24, 311.
Sutherland, R. B., Schuerch, B. M., Ball, S. F. & Horgen, P. A. (1976). Arch. Microbiol. 109, 289.

Takai, N., Uchida, T. & Egami, F. (1967). J. Japan Biochem. Soc. (Seikagaku) 39, 285.

Tashiro, F., Mita, T. & Higashinakagawa, T. (1976). Eur. J. Biochem. 65, 123.

Timberlake, W. E., McDowell, L., Cheney, J. & Griffin, D. H. (1973). J. Bacteriol. 116, 67.

Trapman, J. & Planta, R. J. (1975). Biochim. et Biophys. Acta 414, 115.

Trapman, J. & Planta, R. J. (1976). Biochim. et Biophys. Acta 442, 265.

Trapman, J., Planta, R. J. & Raue, H. A. (1976). Biochim. et Biophys. Acta 442, 275.

Tsuboi, M. (1976). Arch. Microbiol. 111, 13.

Tsuboi, M. & Yanagishima, N. (1976). Arch. Microbiol. 108, 149.

Tsutsumi, K., Majima, R. & Shimura, K. (1974). J. Biochem. 76, 1143.

Uchida, T. & Egami, F. (1971). In: The Enzymes (Boyer, P. D., ed.) Academic Press, New York, London, vol. 4, 205.

Ursino, D. J., Sturani, E. & Alberghina, F. A. M. (1969). Biochim. et Biophys. Acta 179, 500.

Van Assche, J. A. & Carlier, A. R. (1973). Arch. Microbiol. 93, 129.

Van Etten, J. L., Dunkle, L. D. & Knight, R. H. (1976). In: The fungal spore (Weber, D. J. & Hess, W. M., eds.) Wiley, New York, London, Sydney, Toronto 243.

Vasu, S., Hulea, St. & Radulescu, S. (1971). Rev. Roum. Biochem. 8, 351.

Vasu, S., Hulea, St. & Lazar, V. (1972). Rev-Biochem. 9, 179.

Weinberg, R. A. (1973). Ann. Rev. Biochem. 42, 329.

Wejksnora, P. & Haber, J. E. (1976). J. Bacteriol. 127, 128.

Winicov, I. & Perry, R. P. (1974). Biochemistry 13, 2908.

Wither, M. & Stevens, L. (1976). Biochem. Soc. Transaction 4, 1126.

Woodroof, E. A. & Glitz, G. D. (1971). Biochemistry 10, 1532.

Wyers, F., Huet, J., Sentenac, A. & Fromageot, P. (1976a). Eur. J. Biochem. 69, 385.

Wyers, F., Sentenac, A. & Fromageot, P. (1976b). Eur. J. Biochem. 69, 377.

Yanagawa, H., Ogawa, Y. & Egami, F. (1966). J. Biochem. 80, 891.

Yanagida, M., Uchida, T. & Egami, F. (1964). Agr. Biol. Chem. 38, 531.

THE RELATIONSHIPS BETWEEN PIGMENT PRODUCTION AND SPORULATION IN MONASCUS

D. Shepherd

INTRODUCTION

Went (1895) was the first to isolate the fungus responsible for the coloration of "Ang-Khak", also known as chinese red rice, which he named Monascus purpureus. It has been used in the far east for centuries to colour different food products such as fish and wine. It is produced by cultivating Monascus purpureus on cooked rice until the mycelium has totally penetrated the grains, which are then dried and ground into a fine powder, which is used directly. It was not until the 1950's that a detailed study was made on the pigments produced by Monascus (Fielding, 1960; 1961; Haws, 1959) and very little work has been done on their production in submerged, shaken culture (Lin, 1973). Fig. 1 shows the pigments which have been isolated from Monascus. It is known

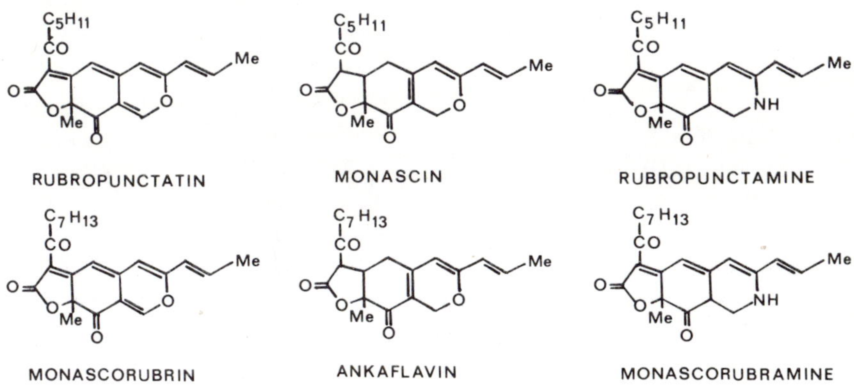

Fig. 1. Pigments isolated from Monascus.

that Monascorubramine and Rubropunctamine (Nakanishi, 1959)

are formed from Monascorubrin and Rubropunctatin by the replacement of an oxygen by an NH-group. The possible relationship between the pigments and their biosynthesis in shaken submerged culture will be discussed later.

SCREENING FOR PIGMENT PRODUCTION

The purpose of this study on Monascus was to select high pigment producing strains, for the production of food co-

TABLE 1

Strains of Monascus Used for Pigment Screening

	ATCC Number
Monascus albidus	16357
M. albidus var. glaber	16358
M. anka	16360
M. anka var. rubellus	16379
M. araneous	16361
M. barkeri	16966
M. fuliginosus	16378
M. major	16362
M. paxii	22080
M. pilosus	16363
M. pubigereus	16364
M. purpureus	6405, 16365, 16385, 16386, 16426, 16427
M. ruber	13692, 15670, 16242, 16366, 16384, 18199
M. rubiginosus	16367
M. rubropunctatus	16368
M. serorubescens	16369
M. vitreus	16371
M. species	16370, 16433, 16434, 16435, 16436, 16437, 16665, 16666, 16771, 16772, 16773, 16774, 16775

lours. A screen was made of 40 different strains of Monascus, all obtained from the American Type Culture Collection. Table 1 shows the 40 strains which were screened. Table 2 shows the different media and the cultural conditions used during this study. All inocula were prepared on the rich MP I medium

TABLE 2

Cultivation Media Used

MP I:	
Glucose	4 %
KH_2PO_4	0.3 %
Yeast extract	1.0 %
pH after sterilization 5.6	
MP II:	
Glucose	4 %
KH_2PO_4	0.1 %
$MgSO_4 \cdot 7H_2O$	0.05 %
NaCl	0.05 %
$FeSO_4 \cdot 7H_2O$	0.01 %
$NaNO_3$	0.3 %
pH after sterilization 5.0	
MP III:	
Glucose	4 %
KH_2PO_4	0.1 %
$MgSO_4 \cdot 7H_2O$	0.05 %
NaCl	0.05 %
$FeSO_4 \cdot 7H_2O$	0.01 %
NH_4Cl	0.05 %
pH after sterilization 5.2	
Temperature:	25°C
Agitation speed:	150 rpm

in order to produce an actively growing mycelium, before transfer into the production media containing nitrate and ammonium as nitrogen source. At the same time as the screen for pigment production, a microscopic examination of the cultures was made in order to study the effect of the different media on sporulation. The usual parameters were measured during the experiments, as well as a quantitative estimation of the number of conidia and different stages of the sexual reproductive cycle. A previous study (Carels & Shepherd, 1975) has shown that all the different stages of the sexual reproductive cycle can be observed in shaken culture.

After our initial experiments it was clear that all the strains could be classified into six groups which are shown in Table 3. Only the presence or absence of a character was used for this classification, no account has been taken of the quantity produced. However, a strain was considered to be a pigment producer if an extract of one millilitre of a 168 hour culture gave an optical density of more than 0.1 at maximum absorption. It can be seen that most strains do not produce pigment but form conidia. There were strains which had a coloured mycelium but not enough pigment was produced to give a coloured extract under standard conditions. It is interesting to see that there are no strains which reproduce sexually and do not form conidia at the same time.

TABLE 3

Classification of _Monascus_ Strains According to Pigment Production and Reproduction System

Pigment	+				–			
Sexual structures	+		–		+		–	
Conidia	+	–	+	–	+	–	+	–
Number of strains	4		3	2	9		17	5

After our first screen on the different media several general observations could be made concerning pigment production and sporulation which are shown in Table 4. These observations were valid for all the strains of _Monascus_ studied.

It has already been shown by Bu'Lock (1961), that balanced but limited growth, as well as nitrogen exhaustion is required for the induction of secondary metabolites. Our results showed that this was also true for the pigment producing strains of _Monascus_. The work already carried out on the

TABLE 4

Observations After First Screen of 40 Strains	
1.	pH evolves differently depending on the nitrogen source
2.	Different pH determined the colour of the pigment formed
3.	Conidia formed (by conidiating strains) on all media early in fermentation
4.	More conidia formed on nitrate medium
5.	Conidiation inhibited on ammonium medium
6.	More pigment produced on ammonium medium
7.	Sexual reproduction develops better on nitrate medium
8.	Sexual strains use very little nitrate during early hours of cultivation
9.	Sexual reproduction reduced on ammonia medium
10.	Conidiation and sexual reproduction occurs on medium containing organic nitrogen

chemistry of the pigments suggests that only two types of pigment, the Monascorubrin/Rubropunctatin type and the Monascin/Ankaflavin type are formed biosynthetically. From their structures it is known that Monascin and Ankaflavin cannot react with NH-groups to produce the corresponding amine, therefore it is only Monascorubrin and Rubropunctatin which form red pigments.

EFFECTS OF CULTIVATION MEDIA

Our results show quite clearly that the cultural conditions determine the type of pigment formed as can be seen from Table 5. On MP I medium only red pigments are produced due to the presence of free amino acids. When either monascorubrin or rubropunctatin are formed on this medium, which has a relatively high pH, they are immediately converted into their respective amines. On the nitrate medium red and orange pigments are found in all extracts. This medium contains no free amino acids, so although the pH is relatively high the pigment can only react with the amino acids of the mycelial pool or other NH-containing compounds in the mycelium. An orange pigment is therefore present. It has been shown that if

TABLE 5

Types of Pigment Produced According to Media Used for Cultivation

Strains	Wavelength nm	MP I				MP II				MP III			
		24	72	120	168	24	72	120	168	24	72	120	168
M. albidus 16357	400												
	470												
	520												
M. major 16362	400					0.1	0.67	1.17	1.33	0.3	1.04	1.95	2.5
	470									0.28	1.44	2.65	3.5
	520					0.1	0.86	1.53	1.54				
M. pilosus 16363	400												
	470												
	520												
M. purpureus 16427	400		0.19	0.31	0.35	0.19	0.94	1.54	1.66	0.22	1.2	2.2	2.48
	470									0.18	1.8	2.85	3.0
	520		0.12	0.23	0.26	0.24	1.12	1.84	1.56				
M. rubiginosus 16367	400					0.12	0.97	1.4	1.1	0.13	1.85		
	470										2.37	4.5	5.6
	520					0.08	0.69	0.59	0.62				
M. species 16436	400		0.37	0.52	0.71								
	470												
	520		0.35	0.56	0.71								

The figures given above are the optical densities at different wavelengths of alcoholic solutions (20 ml) obtained by extracting one millilitre of homogenized culture broth.

ammonia, for example, is added to the cultures they become deep red in colour and no orange pigment can be extracted.

The conditions are totally different when ammonia is in the medium. The pH falls quickly and stays at about 2.5. At this low pH the pigment produced cannot react with any of the NH-groups of the mycelium so monascorubrin and rubropunctatin accumulate.

A yellow pigment is also present in extracts of some cultures grown on ammonia as nitrogen source. Spectra of the isolated pigment suggest it is monascin or ankaflavin. It is known that these pigments can be formed from monascorubrin or rubropunctatin by oxidation with hydrogen peroxide. Kurano _et al_. (1963) have observed that the quantity of monascin increases with the age of the culture. They suggested that monascorubrin is converted into monascin because of an increase in pH due to lysis of the mycelium. In our experiments the yellow pigment appears late in the fermentation but no increase in pH is observed. However, strains which form the yellow pigment form less mycelium, so it is possible that there is more oxygen available in these cultures which could oxidise the orange pigments.

More work is under way to look into the formation of the pigments, because it is possible that the different strains of _Monascus_ produce only monascorubrin and rubropunctatin biosynthetically and the others are formed chemically depending on the cultural conditions.

The cultural conditions also play a role in determining the type of spores formed. It has been shown for other fungi (Morton, 1961)that there is a relationship between the exhaustion of certain nutrients from the medium and the onset of sporulation, as for the onset of secondary metabolism. This does not seem to be the case for _Monascus_, as conidia are formed early when glucose and nitrogen are still largely available.

When organic nitrogen is present in the medium conidia are formed and sexual reproduction can be induced but the latter is much reduced which favours more conidia being formed. This would suggest that yeast extract contains substances which stimulate conidiation, these same substances being able to be synthesized on the other media during the early hours of cultivation. These substances could be amino acids which are the main nitrogen source in the MP I medium. It is known that amino acid synthesis takes place during the early hours of cultivation (Bu'Lock, 1965) and Dicker _et al_. (1969) have demonstrated that the addition of glycine to the medium stimulates conidiation.

As already mentioned the presence of nitrate stimulates conidiation and sexual reproduction while ammonia reduces the number of conidia and sexual structures. The uptake of nitrate indicates that there is a relationship between the efficiency of nitrogen uptake and sporulation. After 8 hours of cultivation most nitrate is used by the non-sporulating strains, less is taken up by the conidiating strains while practically no uptake occurs with the sexual strains.

This does not explain why there are more conidia on nitrate medium and less on ammonium. The efficiency with which the different glycolytic pathways function could be an important factor. The hexose monophosphate pathway is the major pathway involved in biosynthetic reactions and is also the main source of NADPH, a co-factor for nitrate reduction (Pateman & Cove, 1967). This system regenerates NADP, needed for glucose-6-phosphate dehydrogenase and so ensures the functioning of the hexose monophosphate pathway (Fuscaldo et al., 1971). NADPH is also required for the formation of glutamic acid from alpha-ketoglutovate (Kingham & Pateman, 1973). Although the growth potential is high on the nitrate medium if there is any limit in the uptake of nitrate the functioning of the hexose monophosphate pathway will be reduced. If glucose is metabolised by any other pathway there will not be the same formation of NADPH.

The sexual strains have the best initial glucose uptake but the lowest nitrate uptake. This drastically limits biomass formation which is probably why sexual reproduction is induced. However, it would appear that sufficient amino acids are formed, probably by protein breakdown, to insure some conidiation. Strains which sporulate abundantly probably have some limitation on the uptake of nitrate which allows formation of amino acids but not sufficient other intermediates to insure maximum biomass formation hence no conidia are formed.

On the ammonium medium no NADPH is required for nitrogen uptake as it enters the cells by diffusion (Kingham & Pateman, 1973). Nitrogen uptake is therefore not limited by the functioning of the hexose monophosphate pathway. Even if this pathway functions badly, growth can occur but the amount of amino acids will be reduced so reducing conidiation but inhibiting sexual reproduction. These explanations are all very well but one great difference between the two media which should not be overlooked is the pH. When ammonia is present the pH drops sharply to 2.5, whereas on nitrate medium the pH rises during cultivation. It could be that this low pH is the cause of the inhibition of conidiation.

It has already been mentioned that the pH determines the type of pigment produced but it would also seem to determine the amount. There is a relationship between the quantity of pigment produced and the number of conidia formed, because maximum pigment production occurs when conidiation is reduced. Also on nitrate medium non-conidiating strains produce more pigment than conidiating strains. It could be that similar precursors are required for pigment production and conidiation, so if at a low pH conidiation is inhibited, more intermediates can be channelled into pigment production.

EFFECTS OF AMINO ACIDS AND pH

After our initial study with 40 strains, it was clear that conidiation was stimulated by primary metabolites, probably amino acids, and that a low pH inhibited conidiation but stimulated pigment production. This relationship between conidiation and pigment production was obviously very important for the industrialization of the process for pigment production.

With a limited number of strains, shown in Table 6, we studied the effect of amino acids and pH on conidiation and pigment production in more detail.

TABLE 6

Pigment Production and Sporulation of Some *Monascus* Strains

1. *Monascus albidus* 16357: never produces pigment but is capable of conidiation
2. *M. major* 16362: never sporulates but produces pigment
3. *M. purpureus* 16427: forms conidia, reproduces sexually and produces pigment
4. *M. rubiginosus* 16367: forms conidia and produces pigment
5. *M. species* 16436: reproduces sexually, forms conidia and does not produce extractable pigment

Because conidia were formed on the organic nitrogen medium, yeast extract was added to the nitrate and ammonium media. Although a slight increase in conidiation was observed with some strains, the main effect was an increase in biomass, probably due to the vitamins in the yeast extract.

All the amino acids contained in yeast extract were added to the medium, in the same concentration as they are found in yeast extract. These are shown in Table 7. Table 8 shows the results of this experiment. It can be seen that there was a dramatic increase in sporulation and a decrease in pigment production even with the non-sporulating strains. Although not shown in Table 8, there was no appreciable increase in growth. The addition of the different amino acids separately did not have any effect.

TABLE 7

Amino Acids Supplied to the Cultivation Media (mg/100 ml)

Lysine	25.0
Histidine	7.4
Arginine	15.2
Aspartic acid	23.9
Threonine	19.9
Serine	24.9
Glutamic acid	56.9
Proline	9.9
Glycine	12.9
Alanine	34.9
Valine	26.9
Methionine	6.9
Iso-leucine	20.9
Leucine	33.9
Tyrosine	7.9
Phenylalanine	20.9

The effect of pH on sporulation and pigment production was also studied. The pH was regulated to 2.5 at the beginning of the experiment. These results are shown in Table 9. On the nitrate medium no spores were formed early in fermen-

TABLE 8

Influence of Amino Acids on Conidiation and Pigmentation

STRAINS	CONTROL						+ AMINO ACIDS					
	MP 1		MP 2		MP 3		MP 1		MP 2		MP 3	
	CONIDIA	PIGMENT	CONIDIA	PIGMENT	CONIDIA	PIGMENT	CONIDIA	PIGMENT	CONIDIA	PIGMENT	CONIDIA	PIGMENT
16357	4	-	7	-	2	-	5	-	7	-	7	-
16362	-	-	-	2.2	-	2.08	-	-	5	1.04	-	1.7
16367	3	-	2	1.62	2	3.7	3	-	5	0.33	3	0.26
16427	6	-	7	0.2	6	0.4	7	-	7	0.13	6	0.3
16436	5	-	3	-	3	-	5	-	6	-	5	-

TABLE 9

Influence of pH on Conidiation and Pigmentation

STRAINS	CONTROL						pH ADJUSTED TO 2.5					
	MP 1		MP 2		MP 3		MP 1		MP 2		MP 3	
	CONIDIA[a]	PIGMENT	CONIDIA	PIGMENT	CONIDIA	PIGMENT	CONIDIA	PIGMENT	CONIDIA	PIGMENT	CONIDIA	PIGMENT
16357	4	-	7	-	2	-	0	-	5	-	1	-
16362	-	-	0	2.2	0	2.1	0	1.3	0	3.3	0	2.5
16367	3	-	3	1.6	2	3.7	2	0.5	2	2.3	1	3.6
16427	6	-	7	0.2	6	0.43	4	-	6	0.3	5	0.56
16436	5	-	3	-	3	-	4	-	1	-	1	-

a: The numbers of conidia were estimated according to the following code: 0 = no conidia, 1 = 1/4 the average, 2 = 1/2 the average, 3 = average number of conidia, 4 = 1 to 2 times the average, 6 = 4 times the average, 7 = more than 4 times the average.
(The number of conidia is considered to be average when they can be observed under the microscope in any field of view of any sample taken from a culture).

TABLE 10

Influence of pH Adjustment During Cultivation in Nitrate Medium on Pigmentation and Conidiation

		MP 2 (NITRATE)											
		CONTROL			pH ADJUSTED 2.5 OHR			pH ADJUSTED 2.5 24 HRS			pH ADJUSTED 2.5 48 HRS		
HOURS		pH	CONIDIA	PIGMENT	pH	CONIDIA	PIGMENT	pH	CONIDIA	PIGMENT	pH	CONIDIA	PIGMENT
16362	0	5.3	0	-	2.7	0	-	2.7	0	-	2.7	0	-
	24	6.0	0	0.2	5.5	0	0.4	5.5	0	0.4	5.5	0	0.4
	48	6.3	0	1.1	5.6	0	1.7	2.9	0	1.7	5.6	0	1.7
	72	6.3	0	1.7	5.4	0	2.7	3.6	0	2.1	2.9	0	2.1
	96	6.5	0	2.1	6.6	0	2.3	3.1	0	4.0	3.3	0	3.8
16367	0	5.3	0	-	2.7	0	-	2.7	0	-	2.7	0	-
	24	6.1	4	0.1	5.4	2	0.2	5.4	2	0.2	5.4	2	0.2
	48	6.4	3	1.0	6.3	2	1.5	4.6	1	1.5	6.3	2	1.5
	72	6.5	3	1.6	6.8	3	1.8	6.4	3	1.8	4.9	2	2.4
	96	7.0	4	1.3	6.9	4	1.2	7.3	4	1.2	5.0	3	2.8

See Table 9 for explanation of code for quantity of conidia.

tation, but as the pH began to rise conidia appeared but in reduced numbers. The data shown in Table 9 were taken at 72 hours. On the ammonia medium where the pH normally drops quickly the effect was not as marked at 72 hours. However at the beginning of the fermentation no conidia were seen with any of the strains.

Table 10 shows the results obtained from the last experiment which will be discussed. Using the nitrate medium, where the pH rises during the fermentation, the pH was adjusted to 2.5 at 0, 24 and 48 hours. It can be seen that even with a non-sporulating strain like 16362, the pigment is increased when the pH is kept low. With the sporulating strain 16367, when the pH is adjusted a distinct decrease in the quantity of conidia is observed and an increase in the quantity of pigment. If the pH rises again as when the pH was adjusted at 0 or 24 hours, the number of conidia increases and the pigment decreases.

These experiments have allowed us to develop a very simple technique for obtaining maximum pigment production. If the pH is adjusted at the beginning of the fermentation then kept low by the addition of acid, maximum pigment production is obtained and no conidia are formed. This has been shown to be particularly effective with the nitrate medium.

It would appear that when the pH is high, intermediates of primary metabolism are channeled into conidia formation, even in strains which do not normally form conidia and pigment production is reduced. If the pH is low, however, the formation of conidia is inhibited, perhaps certain intermediates are not formed. It is interesting to see that the effect of the pH is reversible and the quantity of conidia or pigment can be regulated by changing the pH.

CONCLUSION

In conclusion it is quite clear that in Monascus we have a very interesting fungus which has the possibility to reproduce asexually and sexually as well as produce secondary metabolites in shaken culture, which allows the study of well phased cultures. Sporulation and pigment production can also easily be regulated and we are now using this simple technique to see if in fact there are direct metabolic links between the formation of conidia and the production of pigments and to see which pathways are stimulated or inhibited by pH.

REFERENCES

Bu'Lock, J. D. (1961). Secondary metabolism and antibiotic synthesis. Advan. Appl. Microbiol. 3, 293-342.

Bu'Lock, J. D. *et al*. (1965). Metabolic development and secondary biosynthesis in *Penicillium urticae*. Can. J. Microbiol. 11, 765-778.

Carels, M. & Shepherd, D. (1975). Sexual reproductive cycle of *Monascus* in submerged shaken culture. J. Bact. 122 No. 1, 288-294.

Dicker, J. W. *et al*. (1969). Amino acid induction of conidiation and morphological alterations in wild type and morphological mutants of *Neurospora crassa*. Arch. Microbiol. 65, 241-257.

Fielding, B. C. *et al*. (1960). Monascorubrin. Tetrahedron Letters, No. 5, 24-27.

Fielding, B. C. *et al*. (1961). The chemistry of fungi. Part XXXIX. The structure of Monascin. J. Chem. Soc. 4579-4589.

Fuscaldo, K. E. *et al*. (1971). Genetic and biochemical studies of the hexose monophosphate shunt in *Neurospora crassa* I. The influence of genetic defects in the pathway on colonial morphology. Can. J. Microbiol. 17, 783-788.

Haws, E. J. *et al*. (1959). The chemistry of fungi. Part XXXVII. The structure of Rubropunctatin. J. Chem. Soc. 3598-3610.

Kingham, J. R. & Pateman, J. A. (1973). NAD and NADP L-glutanate dehydrogenase activity and ammonium regulation in *Aspergillus nidulans*. J. Gen. Microbiol. 78, 39-46.

Kurano, M. *et al*. (1963). Biosynthesis of Monascorubrin and Monascoflavin. Chem. Pharm. Bull. (Tokyo) 11, 358-362.

Lin, C-F. (1973). Isolation and cultural conditions of *Monascus* sp. for the production of pigment in a submerged culture. J. Ferment. Technol. 51, No. 6, 407-414.

MacMillan, A. (1954). The entry of ammonia into fungal cells. J. Exptl. Bot. 7, 113-126.

Morton, A. G. (1961). The induction of sporulation in mould fungi. Proc. Roy. Soc. (London), Ser. B. 153, 548-569.

Nakanishi, K. et al. (1959). Monascorubrin II. Structures of Monascorubrin and Monascanine. J. Am. Chem. Soc. 81, 6339-6340.

Pateman, J. A. & Cove, D. J. (1967). Regulation of nitrate reduction in *Aspergillus nidulans*. Nature (London) 215, 1234-1237.

Went, F. A. F. C. (1895). *Monascus purpureus*, le champignon de l'Ang-Quac, une nouvelle Thélébolée. An. Sci. Nat. Bot. Ser. VIII, 1, 1-18.

MODELLING OF MICROBIOLOGICAL PROCESSES

B. Sikyta, M. Novák and P. Doberský

INTRODUCTION

Even though attempts at an exact mathematical description of microbiological processes date far into the past, only recently have microbiologists acquired sufficient knowledge of the processes in living matter, the technical equipment for their monitoring, and, last but not least, the mathematical apparatus and computing technique necessary for the processing of experimental data, simulation, and modelling. The approach to the modelling of microbiological processes has evolved gradually, undergoing two distinct phases. The first phase, lasting up to the beginning of the sixties, witnessed the development of simple models based on the hypothesis of a homogeneous population; the mathematical equipment required analytical - integral - solutions of the models. The second phase, since the beginning of the sixties until now, has marked the appearance of new aspects. The population is taken to be inhomogeneous - structured - and a number of the models consist of differential equations which can be solved only numerically with the aid of a computer. Further trends include the development of more generalized models describing satisfactorily the course of the actual processes such as the individual phases of a batch cultivation, continuous cultivation, and transient phenomena.

A model employed in the simulation of microbiological processes is a set of mathematical formulae describing the system under study. In general, any system providing information about the processes under study represents a model. The basic variable in any generally valid model describing the dynamics of a cell population is the number of living cells (N) in a given volume. These models are called segregated ones. In practice, the number of cells can often be replaced by the concentration of live cell matter (X) referred to e.g. as the dry weight, optical density (absorbance), etc. These models are called distributed ones since they rely on a basic assumption that the distribution of the cell matter within the given volume is uniform. Distributed models describe changes in the concentration of cell matter such as growth, segregated models describe changes in the number of individual cells, i.e. multiplication.

Each of these groups can be further classified to structured and non-structured models. The latter describe a homogeneous population of cells which are entirely uniform with respect to both cell composition and the growth rate. The more realistic structured models take into account the presence, in the population, of groups of cells with different chemical composition (distributed models) or different age or size (segregated models).

The solution of all these types of models can be either deterministic or stochastic. Deterministic models include as variables mean values and the density of the population or the concentration of the nutrients or products are determined by initial conditions. Stochastic models for corresponding mean values yield equations identical with those obtained with the deterministic models. In addition, however, they make it possible to study the fluctuations of the values around the means.

An important practical classification criterion is the purpose for which the model is set up. The models can be thus divided into

a) models for process control

b) models for simulation.

With models (a) the structure, i.e. the correspondence of the model with the structure of the system under investigation, is of secondary importance, the main stress being laid on the analogy of the behaviour of the model and the studied system. Models of this type have been thoroughly described and discussed (e.g. Ryu & Humphrey, 1973; Nyiri, Jefferis & Humphrey, 1974; Jefferis, 1975); they will not be dealt with here though they represent the basis for the computer control of microbiological processes and their optimization.

The subject of our communication is rather the second group. Let us consider the suitability, main features, and practical value of simulation of microbiological processes. In this respect, modelling is used mainly

a) in experiment planning - the determination of limit experimental conditions,

b) in interpolation and extrapolation of experimental results,

c) for elucidating the nature of the process, its biological, chemical or physical basis.

The first category encompasses e.g. a number of simple models used for the prediction of the behaviour of continuous systems. Abundant literary data exist on the subject (e.g. Luedeking & Piret, 1959; Slezák & Sikyta, 1964; Fencl & Novák, 1969).

The second category includes more sophisticated models providing information on a microbiological process with the range of experimental conditions (interpolation), mainly for economical reasons (saving of time, labour, expensive substrates, etc.), and models yielding information about the process beyond currently used experimental conditions (extrapolation); the reasons for extrapolation are e.g. extremely long duration of the process, insufficient analytical equipment, inability to monitor extremely low concentrations, etc.

The third category represents actual modelling in the strictest sense of the word since models of this class serve to substantiate a hypothesis on the mechanism and function of a given microbiological process and thereby elucidate its nature. The construction of this type of a mathematical model is a cyclic process in which the model is constantly improved by comparing it with the corresponding modelled system. A necessary and, to a considerable extent, crucial prerequisite for model construction is an intimate knowledge of the behaviour of the process under study. This knowledge is based on an adequate information on the time course of both input and output variables, obtained through measurements. The measurement is thus an essential factor in the identification of the behaviour of the process; inaccurate measurements may become a limiting factor in the model build-up, preventing the formation of precise conclusions.

The quality of measurement is a necessary but not satisfactory condition of a biological experiment. An equally important and sometimes crucial factor is the scatter of system parameters (e.g. among individuals of the same species) and time changes in these parameters. Another source of difficulties may be the impossibility to isolate the process from random fluctuations in the environment which can affect the process behaviour.

Extremely valuable in this context are automated systems for data retrieval and processing, which are highly effective in gathering reliable data pertaining to a process. The precise measurement represents but the first stage, to be followed by efficient data filtration, statistical processing and the transformation of the data to the most informative form. Both the data processing preceding the modelling and the modelling itself are currently carried out with the aid of computing technique.

EVALUATION OF MICROORGANISM GROWTH CURVES

An example of the interpolation application of mathematical models may be the frequently needed determination of an optimum time course of experimental growth curves of microorganisms. All kinetic quantities are known to be related to the biomass concentration or the counts of the microorganism in a certain volume; these values, or their time courses - growth curves - serve as the basis for further kinetic analysis. The experimental data are usually hardly suitable for the purpose, exhibiting a scatter and being hampered with a certain error. Some form of data regression is thus needed to obtain suitable results without introducing further inaccuracy into subsequent calculations. The conventional procedure, which may often not be recognized as a regression, is the graphical analysis - the fitting of the typical sigmoid curve to the data. The resulting growth curve yields then "smoothed-out" values for any given time, which represent consequently the best estimate. The original discontinuous function, which can be graphically represented as a broken line, gives thus a continuous function, a curve, with all attendant advantages such as a continuous derivative, etc.

The same result is obtained when a computer is used; the whole operation becomes considerably more accurate and faster. The evaluation is then done by numerical analysis using a suitable regression method, e.g. the least square method. A mathematical model is then used as the regression function, to which the experimental data are related.

In our study, the regression of the growth curve including both the exponential and the stationary growth phase was done by means of a simple model described by Kono (1968). The regression functions were as follows:

Exponential phase

$$X = X_L \exp(\hat{\mu}(t - t_L)) \tag{1}$$

Declination phase

$$X = X_m - (X_m - X_c) \cdot \exp\left(-\frac{\hat{\mu} X_c}{X_m - X_c}(t - t_c)\right). \tag{2}$$

Tentative values of curve constants were determined by approximating the relationship $dX/dt = f(X)$ by linear regression and precise values were obtained by a modified Rosenbrock (1970) method.

Fig. 1 shows the experimental points and, for illustration, the successive approximations of the growth curve approaching the optimum shape. The constants of the model, $\hat{\mu}$, X_c and X_m, t_c, t_L, corresponding to the best-fitting curve shape, make it possible to calculate the value of X for any one moment. In addition, the analytical form of function X(t) facilitates a number of mathematical operations such as differentiation, and can be used for simulation.

As stated above, a mathematical model was used in this case as the regression function. It should be noted, however, that other procedures exist for the approximation of the growth curve by mathematical functions of X(t). The characteristic sigmoid shape can for instance be approximated by the function

$$X = A \cdot t \exp(-t^B) \tag{3}$$

where A and B are constants, or by polynomials of the type

$$X = \frac{K}{1 + \exp(F(t - t_L))} \tag{4}$$

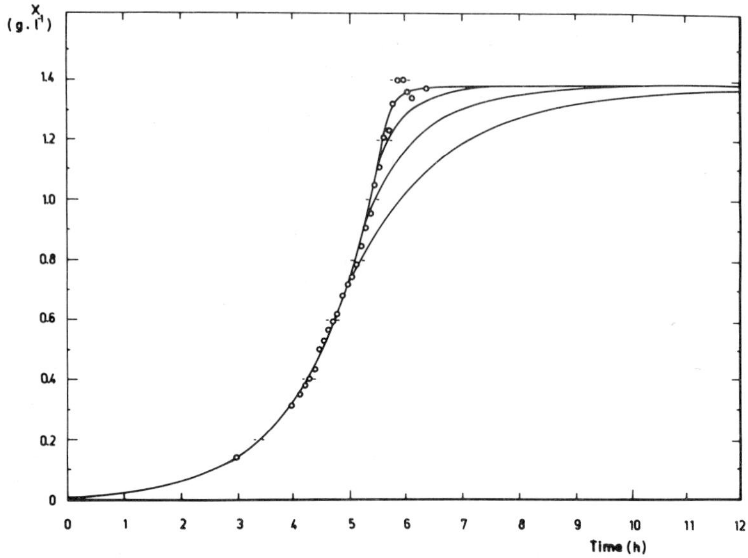

Fig. 1. Regression of the growth curve of E. coli grown in a mineral medium with lactose at 37°C. Lines in the upper part of the growth curve represent sequential approximation. Marks on curves represent respective inflection points (viz. X_c).

where K is a constant and

$$F(t-t_L) = a_o + a_1(t-t_L) + a_2(t-t_L)^2 + \ldots + a_n(t-t_L)^n \qquad (5)$$

where a with indices are constants (Edwards & Wilke, 1968). These functions approximate well the experimental data, can be differentiated, and are also amenable to other mathematical manipulations. Such functions are generally known as logistic functions and are sometimes used instead of mathematical models; they can be also readily used in regression. However, their use has an inherent drawback which will be illustrated on the following example.

Experimental data from the exponential phase (Fig. 2) were fitted with different regression functions, with results summarized in Table 1. Regression coefficients (r^2) for individual functions indicate the accuracy of the fit of the corresponding functions; $r^2=1$ represents an exact agreement of experimental and function-derived values. A comparison

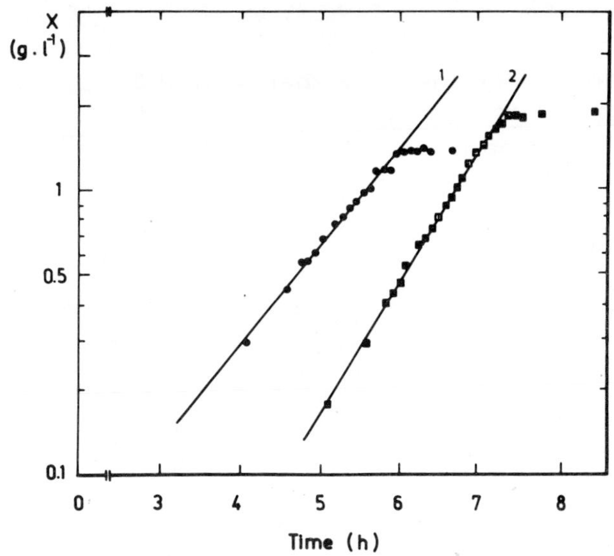

Fig. 2. Fitting of the theoretical curve to experimental data during the exponential phase of growth of E. coli in a mineral medium with lactose at 37°C.
1 = run 1, 2 = run 2.

of these regression coefficients indicates that the best fit is obtained with the function $X = a.\,t^b$. However, the concept of the proliferation of daughter cells by the division of mother ones implies that the increase in the number of microorganisms in the exponential phase is proportional to their number or – in our case – to the biomass concentration

$$dX/dt = bX, \qquad (6)$$

where constant b represents the maximum specific growth rate, $\hat{\mu}$; integration yields then a formula $X = a\ .\ \exp(bt)$ (Table 1). This relation has been thoroughly experimentally verified. On the other hand, differentiation of the equation $X = a\ .\ t^b$ shows that the growth of biomass is proportional to time:

$$dX/dt = ab\ .\ t^{b-1} \qquad (7)$$

which is at variance with the above conclusions.

Thus the equation $X = a\ .\ t^b$ represents a mere logis-

TABLE 1

Regression Functions for Experimental Data Fitting

Regression function	Coefficients a	b	Regression coefficient, r^2	Run
$X = a. \exp(bt)$	0.0125	0.8051	0.9665	1
	0.0008	0.8929	0.9948	2
$X = a + bt$	-2.1284	0.5775	0.9502	1
	-2.5600	0.4314	0.9452	2
$X = a. b^t$	0.0125	2.2369	0.9665	1
	0.0008	2.4421	0.9948	2
$X = a. t^b$	0.0011	4.0024	0.9737	1
	0.0000	6.3016	0.9956	2
$X = a + b\ t$	-4.9649	2.5633	0.9419	1
	-5.5709	2.2814	0.9348	2
$X = a + b/t$	3.5224	-13.6628	0.9059	1
	3.4494	-20.7797	0.8976	2

tic function whilst equation $X = a \cdot \exp(bt)$ is the actual model of a real system. The model is inherently more correct than the logistic function although in our case it fits the experimental results less satisfactorily owing to the scattering of the data.

Generally it can be stated that logistic functions are based on experimental data whereas models predict the data. This distinction, though insignificant at first sight, is essential since it stresses the difference between a mere mathematical description without any biological or physical meaning and an approach based on an intimate knowledge of the laws of growth, multiplication, and metabolic processes in the living matter.

In addition, the above example illustrates still another hazard which can give rise to spurious conclusions, if the selection of the regression function is left to the computer. If the computer itself carries out the identification of the model, the selection criterion being the best fit of experimental data with the function chosen for fitting, then the procedure leads to incorrect conclusions.

MODELLING OF POPULATION CHANGES IN CONTINUOUS CULTURES OF MICROORGANISMS

An important problem in the study of continuous cultivation of microorganisms is population stability. The populations may consist of two or more microbial species (e.g. in ecological studies) and population changes may be caused by mutation with a subsequent selection of one of the species. Small changes in population have no effect on the course of the steady state whereas extensive changes may have profound effects, especially concerning the formation of products.

Undesirable population changes were observed e.g. during the continuous cultivation of Micrococcus glutamicus, a microorganism dependent on methionine and threonine and producing lysine (Slezák, Sikyta, Plachý & Břečka, 1969). The continuous cultivation was carried out in a complex medium at a dilution rate of 0.3 h^{-1}. At the beginning of cultivation, the population consisted of 95 - 96 % auxotrophs and 4 - 5 % prototrophs; after 100 h the number of prototrophs increased markedly to replace completely, during the following 60 h the population of auxotrophs. This resulted in a sharp drop of lysine concentration in the medium.

Similarly, the drop in the production of proteases and amylase during the continuous cultivation of Bacillus subtilis was attributed by Heineken & O'Connor (1972) to genetical changes.

On the other hand, continuous cultivation can be used for a controlled accumulation of mutants with increased activity, such as the mutants synthesizing higher amounts of enzymes under growth limitation caused by specific substrates. The basic requirement is that the mutant should exhibit, under given conditions, a higher specific growth rate than the original culture.

The explanation of the mechanisms of the above changes is usually difficult even though the nature of the process permits sometimes an estimate, or even a control, of the trends in the population changes. The quantification of our concepts, and a restriction of the empirical approach, greatly facilitates the simulation of the process with a computer. Because of the complexity of population changes, the following treatment was restricted to a population comprising, at most, two different microorganisms (mutants), and the description of the population behaviour was done by the Monod model. The principal factors were taken to be the differences in $\hat{\mu}$ and K_s at different dilution rates.

Fig. 3 summarizes different cases of individual dependences of specific growth rate, μ, on the concentration

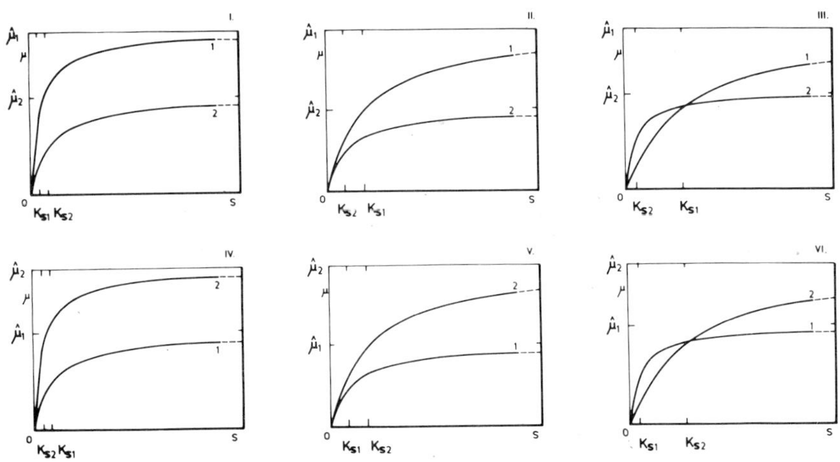

Fig. 3. Theoretical dependences of specific growth rate (μ) on the concentration of a limiting substrate (S) for a population including two microorganisms (mutants).

of the limiting substrate, S. The dependences μ (S) can be divided into two main types depending on the values of $\hat{\mu}$ and K_S of the two microorganisms (mutants). The first type (cases I, II, IV, V) is characterized by the fact that the μ (S) curves for the two microorganisms (mutants) have no common point apart from the origin, in the second type (cases III, IV) the two curves have an intersection. All these cases are illustrated in Fig. 4, where correspondingly denoted areas reflect the magnitude of the ratios $\hat{\mu}_1/\hat{\mu}_2$ and K_{sl}/K_{s2} (indices 1 and 2 refer to the one and the other part

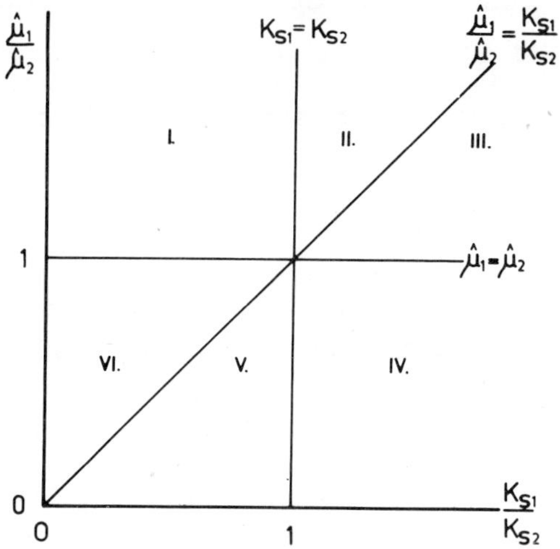

Fig. 4. Generalized plot of functions from Fig. 3. The numbers within areas correspond with those in Fig. 3.

of the population). Thus with known values of μ and K_S for two microorganisms (mutants) Fig. 4 or Table 2 can serve for the identification of a given trend and for the prediction of population changes. The curves with intersections illustrate especially interesting population changes since at low substrate concentration the dominant population component is that with the lower saturation constant K_S while at higher substrate concentrations the prevalence is with the micro-

TABLE 2

Population Changes in Mixed Culture

Type of mixed culture (Fig. 3)	I.	II.	III. $D<\mu_c$	III. $D>\mu_c$	IV.	V.	VI. $D<\mu_c$	VI. $D>\mu_c$
Predominant organism	1	1	2	1	2	2	1	2

organism (mutant) exhibiting the higher maximum growth rate $\hat{\mu}$. The dilution rate can thus be used as a controlling factor and its changes can be made to govern directly the course of population changes.

The simulation of population changes in a continuous system was based on two possible types of behaviour of the culture:

a) at t = O the population is mixed; with proceeding cultivation the ratio of counts of the two microorganisms (mutants) changes solely due to different rates of growth;
b) the composition of the population changes due both to different rates of growth and to a mutation of the parent population (mutation rate λ).

The behaviour of transient states of individual mixed cultures in a continuous system was described by simple differential equations based on the Monod model.

Case (a)

$$\frac{dX_1}{dt} = \hat{\mu}_1 \frac{S}{K_{s1} + S} \cdot X_1 - DX_1 \tag{8}$$

$$\frac{dX_2}{dt} = \hat{\mu}_2 \frac{S}{K_{s2} + S} \cdot X_2 - DX_2 \tag{9}$$

$$\frac{dS}{dt} = -\hat{\mu}_1 \frac{S}{K_{s1}+S} \cdot \frac{X_1}{Y_1} - \hat{\mu}_2 \frac{S}{K_{s2}+S} \cdot \frac{X_2}{Y_2} + D(S_o - S) \tag{10}$$

Case (b)

$$\frac{dX_1}{dt} = \hat{\mu}_1 \frac{S}{K_s + S} \cdot (1 - \lambda) \cdot X_1 - DX_1 \tag{11}$$

$$\frac{dX_2}{dt} = \hat{\mu}_1 \frac{S}{K_{s1} + S} \cdot \lambda \cdot X_1 + \hat{\mu}_2 \frac{S}{K_{s2} + S} \cdot X_2 - DX_2 \tag{12}$$

$$\frac{dS}{dt} = -\hat{\mu}_1 \frac{S}{K_{s1} + S} \cdot \frac{X_1}{Y_1} - \hat{\mu}_2 \frac{S}{K_{s2} + S} \cdot \frac{X_2}{Y_2} + D(S_o - S) \tag{13}$$

Population changes in a continuous culture brought about by a mutation of a part of the population were dealt with previously (Sikyta & Fencl, 1976). With respect to simulation, the mutation rate does not affect appreciably the character of the resulting relationships, its effect being negligible as compared with the effect of other factors. The mutation rate was therefore not included and the description was done using equations (8), (9), and (10).

The role of a simulation in the elucidation of the underlying mechanism of a process will be illustrated by the following example. A continuous cultivation of Streptomyces aureofaciens under sucrose limitation at low dilution rates yields a population which produces neither chlortetracycline nor pigments. The original population was found to be heterogeneous, consisting of two types of cells: production cells, synthesizing both the antibiotic and the pigments, and nonproduction ones, which prevail in a population growing at low dilution rates (Slezák & Sikyta, 1964; Sikyta, 1964). The nonproduction part of the population has a higher growth rate at lower sucrose concentrations, i.e. at low dilution rates ($D < 0.05\ h^{-1}$) and is thus accumulated in the culture. At higher dilution rates ($D > 0.05\ h^{-1}$) the situation is reversed. The population changes can thus be characterized indirectly by changes in the concentration of chlortetracycline and pigments (Fig. 5). The time course of the concentrations of these substances during a continuous cultivation implies that in this case the μ (S) curves have an intersection (cf. Fig. 3, case III, VI). The coordinates of the intersection are defined by equations

$$D_c = \mu_c = \frac{\hat{\mu}_1 K_{s2} - \hat{\mu}_2 K_{s1}}{K_{s2} - K_{s1}} \tag{14}$$

$$S = \frac{\hat{\mu}_1 K_{s2} - \hat{\mu}_2 K_{s1}}{\hat{\mu}_2 - \hat{\mu}_1} \tag{15}$$

To verify if the changes in the concentration of individual components are indeed due to population changes in the culture, the course of the process was simulated for the following experimental values ($\hat{\mu}_1 = 0.15\ h^{-1}$, $\hat{\mu}_2 = 0,12\ h^{-1}$,

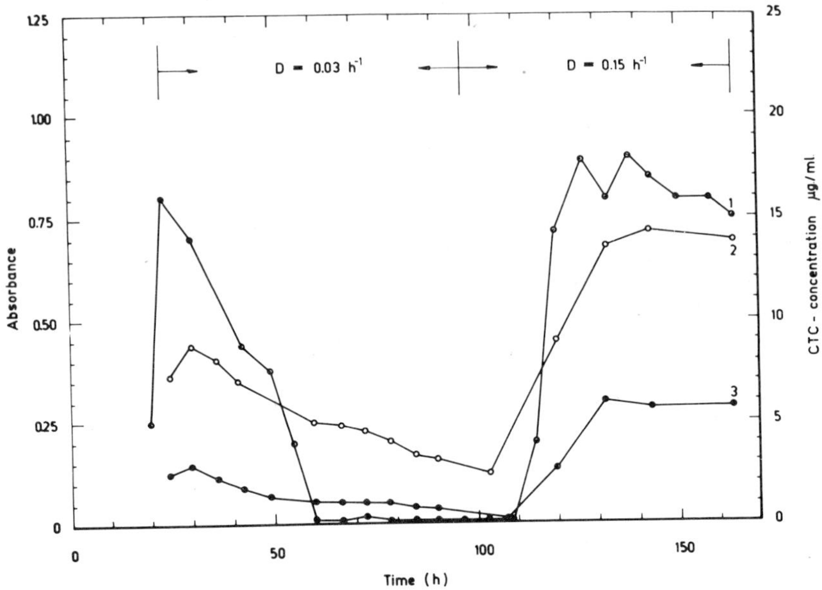

Fig. 5. Absorbance (pH 7) and antibiotic concentration following dilution rate changes as a function of time. Absorbance at 270 nm (O) and at 370 nm (●), chlortetracycline conc. (⊙).

$K_{sl} = 0, 1\ g.l^{-1}$, $K_{s2} = 0.05\ g.l^{-1}$). The critical dilution rate, resulting from these values, was $D_c = 0.05\ h^{-1}$. The time course of the simulated variables is illustrated in Fig. 6. It can be seen that the agreement of the simulated and real time course is fairly good for pigments but not with respect to chlortetracycline. These findings are in keeping with the assumption that the biosynthesis of pigments is associated with growth whereas the biosynthesis of the antibiotic is not. Thus the time course of concentrations of substances associated with culture growth corroborated the original prediction of the action of population changes on the behaviour of the culture.

The simulation of population changes brings another major advantage in that it facilitates a relatively accurate extrapolation of the changes for systems in which the corresponding experiments are difficult to perform owing to prolonged attaining of the steady state. On the other hand,

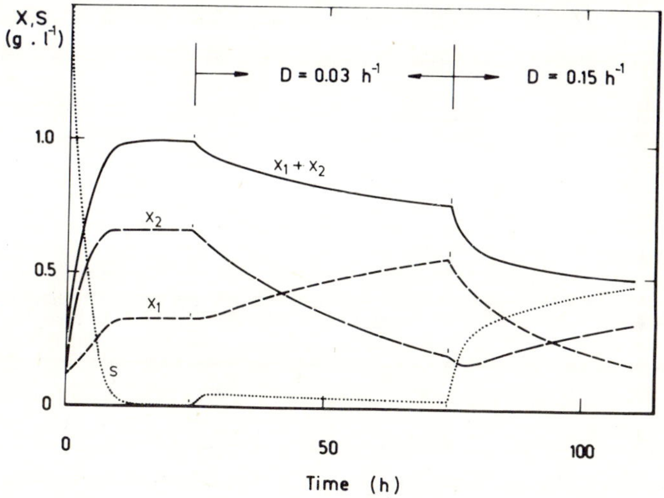

Fig. 6. A simulation of the time course of production and nonproduction part of population of *Streptomyces aureofaciens*; ($\hat{\mu}_1$ = 0. 15 h^{-1}, $\hat{\mu}_2$ = 0. 12 h^{-1}, K_{s1} = 0. 1 g. l^{-1}, K_{s2} = 0. 05 g. l^{-1}).

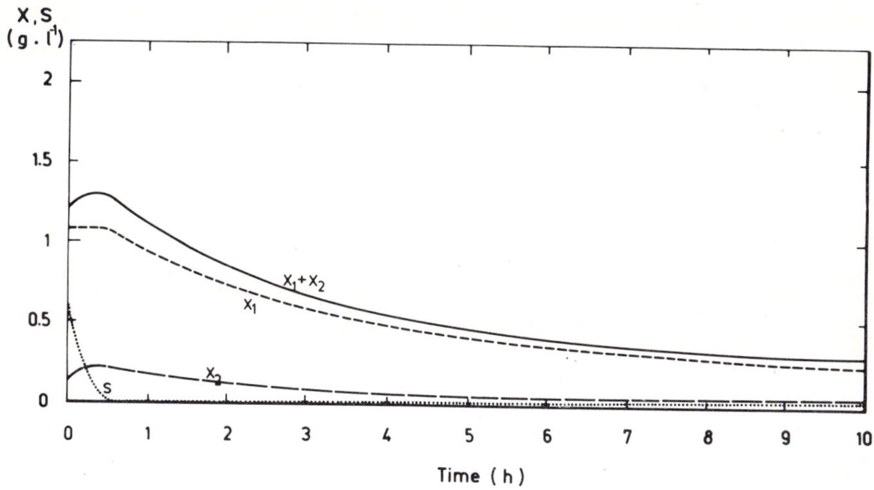

Fig. 7. A simulation of the time course of concentration of microorganisms (X_1, X_2) and a limiting substrate (S) for a short time interval (10 h).

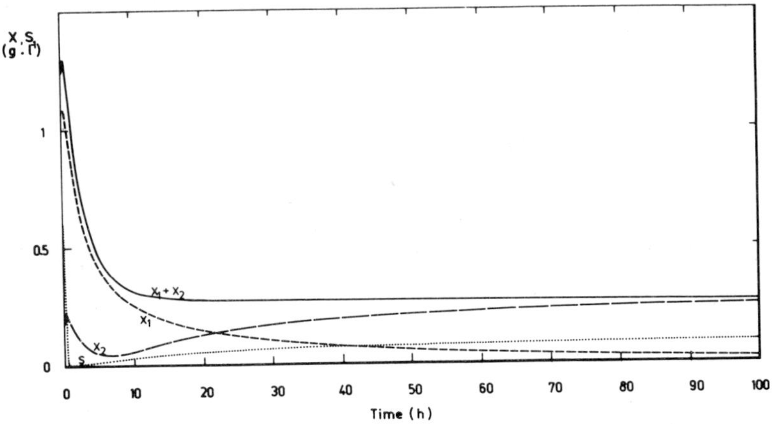

Fig. 8. A simulation of the time course of the concentration of microorganisms (X_1, X_2) and a limiting substrate (S) for a long time interval (100 h).

it should be bornc in mind that an estimate of population changes based on a relatively short experiment may lead to errors (cf. Figs. 7 and 8). Fig. 7 shows the time course of concentration change of two microorganisms (mutants) with different μ and K_s, as determined within 10 h. The marked change in the concentration of the microorganisms resembles superficially an outflow of both of them under given conditions. When the simulation is carried out for a considerably longer time interval (Fig. 8), the outflow is clearly seen to occur with one microorganism only, the other being retained and growing.

These limited examples illustrate the significance of both modelling procedures and computing technique in the study and appraisal of microbiological systems. The modelling is still largely hampered by the fact that the quantitative relationships governing the mechanisms of individual processes under different experimental conditions are not known. On the other hand, this inadequate knowledge does not always restrict the use of the models.

SYMBOLS USED

D dilution rate (h^{-1})

K_s saturation constant ($g. l^{-1}$)

S	concentration of limiting substrate ($g.l^{-1}$)
t	time (h)
t_c	time at the inflex point of the growth curve (h)
t_L	time at the beginning of the exponential phase (h)
X	concentration of biomass ($g.l^{-1}$)
X_c	concentration of biomass at the inflex point of the growth curve ($g.l^{-1}$)
X_L	concentration of biomass at the beginning of the exponential phase ($g.l^{-1}$)
X_m	maximal concentration of biomass ($g.l^{-1}$)
μ	specific growth rate (h^{-1})
$\hat{\mu}$	maximum specific growth rate (h^{-1})
λ	mutation rate

REFERENCES

Edwards, W. H. & Wilke, C. R. (1968). Biotechnol. Bioeng. 10, 205.

Fencl, Z. & Novák, M. (1969). Folia microbiol. 14, 314.

Heineken, F. G. & O'Connor, R. J. (1972). J. gen. Microbiol. 73, 75.

Jefferis, R. P. (1975). Process Biochemistry 10, 2.

Kono, T. (1968). Biotechnol. Bioeng. 10, 105.

Luedeking, R. & Piret, E. L. (1959). J. Biochem. Microbiol. Technol. Eng. 1, 393.

Nyiri, L. K., Jefferis, R. P. & Humphrey, A. E. (1974). Advances in Microbial Engineering, Part 2. Biotechnol. Bioeng. Symp. Nr. 4, 168.

Rosenbrock, H. H. & Storey, C. (1970). Computational Techniques for Chemical Engineers, Pergamon Press, Oxford.

Ryu, D. Y. & Humphrey, A. E. (1973). J. Appl. Chem. Biotechnol. 23, 283.

Sikyta, B. (1964). Continuous cultivation of Streptomyces aureofaciens. Rozpravy ČSAV, řada MVP, 74.

Sikyta, B. & Fencl, Z. Continuous Culture 6 (1976). Applications and New Fields, E. Horwood Ltd., Chichester, p. 158.

Slezák, J. & Sikyta, B. (1964). Continuous cultivation of microorganisms. Proc. 2nd Symp., Academia, Prague, p. 185.

Slezák, J., Sikyta, B., Plachý, J. & Břečka, A. (1969). Continuous cultivation of microorganisms. Proc. 4th Symp., Academia, Prague, p. 505.

NEW FERMENTER CONFIGURATIONS

H. W.-D. Katinger

INTRODUCTION

Designing a fermenter is a wholly interdisciplinary activity. The designer has to make use of well defined mathematical, physical and mechanical regularities in order to meet the poorly defined requirements of a submerged microbial deep culture. The goal of his efforts is to give a fermenter a geometric configuration. The configuration of the fermenter <u>as such</u> serves as the rheologic transmission that has to transform mechanical power into a series of mass transport steps and into liquid bulk flow phenomena, which are necessary to keep a microbial culture growing. Both mass transport and the nature of bulk flow will to a great extent guide the activity and the physiological differentiation exhibited by the culture. In any fermentation process the configuration must therefore play the part of a mediator between power input variables and the physiological reactions of a culture.

Delivering a lecture on "new fermenter configurations" requires one to consider the subject in the context of both sides - biologically on the one, abstract and mechanically on the other. It is quite clear, however, that my illustrations will be quite insufficient exhaustive enough to satisfy in either case the expectations of the specialists concerned. I shall instead try to depict the connective elements between the disciplines involved in fermenter design, and point out some topics that represent problems still unsolved in the operation of many fermentations, in particular those using filamentous microorganisms.

Fermenters are required to perform a great number of functions which may sometimes impose contradictory or

TABLE 1

Desired Properties for Fermenters
Functional requirements:
- High gas/liquid mass transfer
- Creation of gas/liquid interfaces but no foaming problems
- Sufficient hold up of dispersed phases (gaseous liquid or solid)
- Reasonable heat-transfer
- Avoid "dead" zones also in big fermenters and in high-viscous mashes by control of bulk flow
- Prevent aggregation but no damage of microorganisms etc.
Economic requirements:
- Cheap, robust and simple mechanical design
- Easy to operate under aseptic conditions
- Well understood scale up characteristics
- Reasonable flexibility with respect to various process requirements
- Stable operation under fluctuating process condition
- Low specific power consumption

conflicting requirements on the design of the equipment. Generally these requirements can be correlated as shown in Table 1. In particular they can be divided into two groups, one functional and one economic. First of all the fermenter must have sufficient gas/liquid mass transfer in order to prevent limitations of the microbial activity. The creation of high gas/liquid interfaces may often lead to undesirable foaming problems. The method of air sparging is therefore very critical for the inherent defoaming capacity that a fermenter is also required to have.

Another critical feature to be pointed out is the amount of bulk flow generated in the fermenter. A given minimum bulk flow is absolutely necessary in order to avoid dead zones not involved in the overall mixing process of the fermenter. Bulk flow must also provide the hold-up of dispersed phases and for their uniform distribution over the entire fermenter volume.

Fermenters of the so called standard (or "Rushton") configuration usually tend to exhibit drastically reduced bulk flow with increased scale of operation. This undesirable behaviour may very often cause severe problems of inhomogeneity. Heat transfer can be considered as reasonably realized when in addition to the turbulent heat generated in the fermenter, for each mole of oxygen transferred, somewhat more than 0, 11 Mcal are removed by cooling under isotherm operation in all regions of the fermenter (Cooney _et al._, 1969).

Phenomena of aggregation and disaggregation and the damaging effects of stress factors on the cells, as they occur in submerged deep culture, are still quite poorly understood.

The commercial success of fermentation processes is mostly affected by this second group of features. A process would never operate on a competitive basis if these requirements were not fully realized. Most of the selling arguments of the fermenter manufacturers are in this group. Cheap, robust and simple mechanical design will be as welcome from the user's side as high quality of mechanical finish. However, the maintenance of aseptic operation will greatly depend on the latter; i.e. the devil of asepsis lurks in the details of mechanical finish and material.

The other features listed here, such as scale-up characteristics, flexibility with respect to various process requirements and (not least) specific power consumption are of the utmost economic importance and are again closely related with the configuration of a fermenter. Simple, even transparent, configurational conceptions are surely a good prerequisite for attacking these requirements in a predictable manner.

TRENDS AND PROBLEMS IN FERMENTER DESIGN

Let me now attempt to outline how fermenter configurations have developed, and how they have been changed

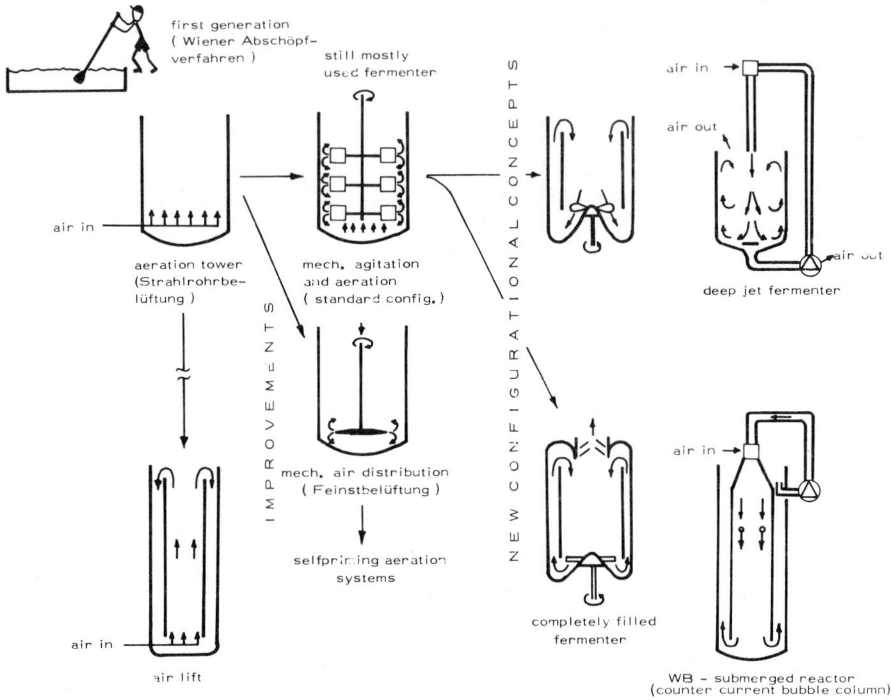

Fig. 1. Development of fermenter configurations (air lift fermenter: Imperial Chemical Industries, German Patent 2135762; WB-submerged reactor: Waagner-Biro AG, German Patent 2627231; deep jet fermenter: Schreier, 1974; completely filled fermenter: Karrer et al., 1976).

over the last decades (Fig. 1). As one may perhaps recognize from this outline (which is far from comprehensive) the fermenters of the first generation are characterized by a very simple and easily understandable configuration and power concept: i.e. overdimensioned petri dishes and man power.

However, in the 20th century, as time became less contemplative and man power became more and more expensive the aeration tower (= Strahlrohr-Belüftung) has been introduced. Further improvements of aeration and of agitation have however been attained by progress along two essentially different routes, either by the installation of mechanical aeration and agitation devices with high power dissipation or by air distributor devices acting with low power consumption. It must be recognized that in practical operation considerably increased fermentation capacities and yields have been attained by the former approach. This is the "standard configuration" and it is still the most frequently used type of fermenter, is found in a great variety of derivative configurations. Countless publications, often very sophisticated, are concerned with the performance characteristics of those fermenters, a fact that in turn reflects the complicated nature these fermenters really have.

On the other hand, with a few exceptions, fermenter configurations for which patents have been given in the last few years have one common feature; they all exhibit forced and controlled liquid bulk flow and belong to the category of so called recycle or loop fermenters, in which the fermenter contents are subjected to a controlled recycle flow either within the fermenter or by an external recycle loop. Stirring is replaced by pumping, either by pneumatic means (as in the case of airlift fermenters) or mechanically achieved.

Thus the very interesting situation which arises is that at the present time the most <u>wide spread</u> fermenter type is still the conventional stirred tank fermenter and the most <u>favoured</u> fermenter type seems to be the bulk flow or recycle fermenter. A second point of interest has also to be remarked: most of these recycle type fermenters were primarily developed by institutions which traditionally have not been involved in the business of fermenter design. New projects of single cell protein production on unconventional C-sources (e.g. n-alkanes, methanol, methane etc.), and in waste water treatment have seemingly stimulated the development of these new fermenter configurations.

In view of these facts it is very difficult to judge the real situation clearly. Let me therefore try to come to a clearer picture from a more abstract point of view. Some considerations on the methods used for tailoring a fermenter configuration and on the most general characteristics of a microbial population in submerged deep culture may help us to elucitate the matter.

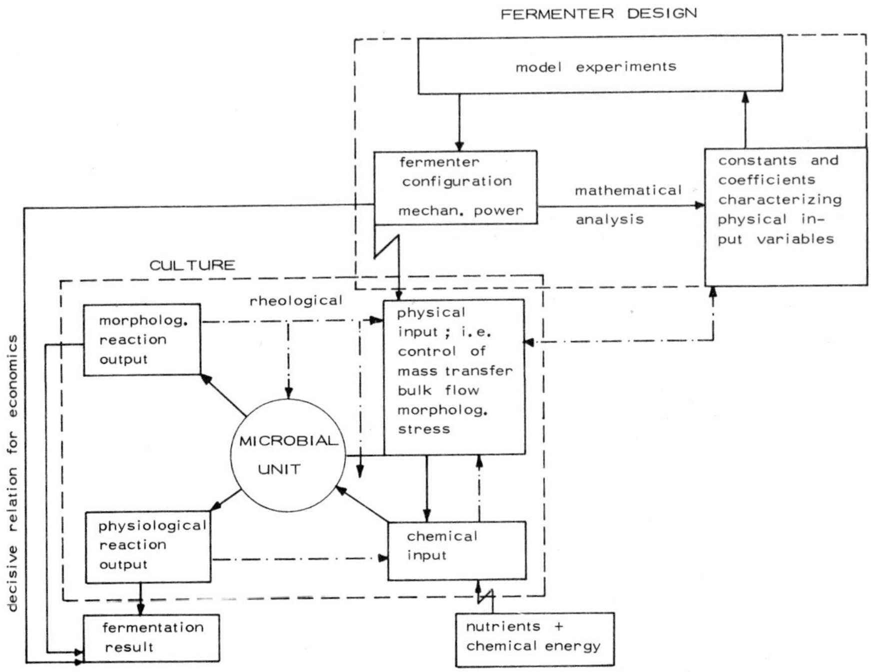

Fig. 2. Fermenter design - the problems' abstraction.

Let me start with a few statements regarding the business of fermenter design. A fermenter is a very complex and complicated machine. It has to distribute the mechanical power, which is supplied from outside, into different physical events which, by their nature, are a system of forces. The nature of these forces can be analyzed and described mathematically (i.e. by dimension analysis). The result of this analysis is expressed in a number of constants and coefficients, which ultimately characterize qualitatively the functions of a fermenter. In order to quantify these coefficients and to coordinate them to a certain configuration, model experiments are required. For reasons of simplicity these experiments are usually performed in pure water or defined aqueous systems.

The fermenter configuration and its power relationships are thus traced out and optimized in a repeated series of model experiments and mathematical analysis. By this way also the constants and coefficients are quantified, and will characterize the nature of the fermenter. The whole business is very laborious and expensive, since the results of the quantification of coefficients and constants may drastically change with the scale of the model experiments. The more complex the configuration of the fermenter is, the more its coefficients and constants will be altered with scale-up (or scale-down). This underlines the complicated similarity relationships of the standard fermenter configuration (i.e. the stirred tank reactor). It is therefore understandable that the preferred basic configurational conceptions of fermenters should be as simple as possible, particularly since the system becomes (by several orders of magnitude) even more complicated when the well-defined model experiments are replaced by a growing microbial culture. Irreducibly, a fermenter must always be considered together with the microbial system. It may sound commonplace to mention this, but the microbes' fate is too often forgotten by modelling enthusiasts.

In the context of fermenter design the microbial culture can be simply considered as a dynamic system whose activity is kept in a dynamic physiological state by the chemical input of nutrients and chemical energy and by the physical input variables that control mass transfer (i.e. physico-chemical input), bulk mixing, morphological stress etc. (Fig. 2). These latter functions are themselves determined by the fermenter configuration and by input mechanical power. The microbe's activity is manifested by a determined physiological reaction output and by the morphological reaction picture. The overall fermentation result is determined by both.

The whole system tends to change its expression as a result of the microbes' own activity. The release of substances into the liquid bulk is feeding back upon the chemical input variables and will thus lead to an altered regulatory expression in the microbial mass. The release of surface active proteinaceous substances will affect the rheological properties of the liquid bulk and thus feed back on air distribution, gas hold up, diffusional mass transfer etc.

Morphologic feed back (e.g. cell aggregation) may even alter the rheology of the entire culture so drastically that physical control over the fermentation is sometimes totally lost. More generally, the rheologic feed back imposed upon the fermenter performance by the microbes' activities will lead to a changed interrelationship between the fermenter functions, i.e. in the constants and coefficients characterizing the fermenter function. In other words, morphologic feed back will shift the gear ratio of our "hydraulic transmission". Fermentation technologists therefore, quite naturally, tend not to have too much confidence in these characteristics. They have much more faith in practical experience gained from direct comparison of different fermenter types with the fermentation results achieved. One must not forget that the environment under which fermentation processes are operated in practice is more an economic one than an academic one. This is also one main reason that will guarantee standard fermenter configurations a long life, because fermentation companies cannot afford to give up much of the well established practical experience which they have with standard fermenters. A body of established "know-how" is preferred to an elusive "know-why". A good chance for new fermenters to become established as quickly as possible is therefore to demonstrate that they are more efficient than standard fermenters in established fermentation processes. This will not be very difficult for fermentations with single-celled organisms. The deviation in fermenter performance between model experiments and the actual fermentation are usually not so critical as those which exist for aggregating and filamentous organisms, where there is strong rheologic feed back.

The problems resulting from cell aggregation are dependent on the nature of the aggregates (Fig. 3). As shown on this outline two different phenomena of aggregation can be differentiated. The first one results from electrostatic interaction and/or protein-protein interaction on the cell surfaces (Shu Chien, 1975; Smidt, 1976). This form of aggregation is easily disaggregated into single cells when the aggregate passes the shear fields of a fermenter. The cells will also, spontaneously, reaggregate and thereby be mixed over in their location within the aggregate. No "ageing" of cells will therefore occur as long as the bulk flow of the fermenter guarantees that the aggregates are periodically

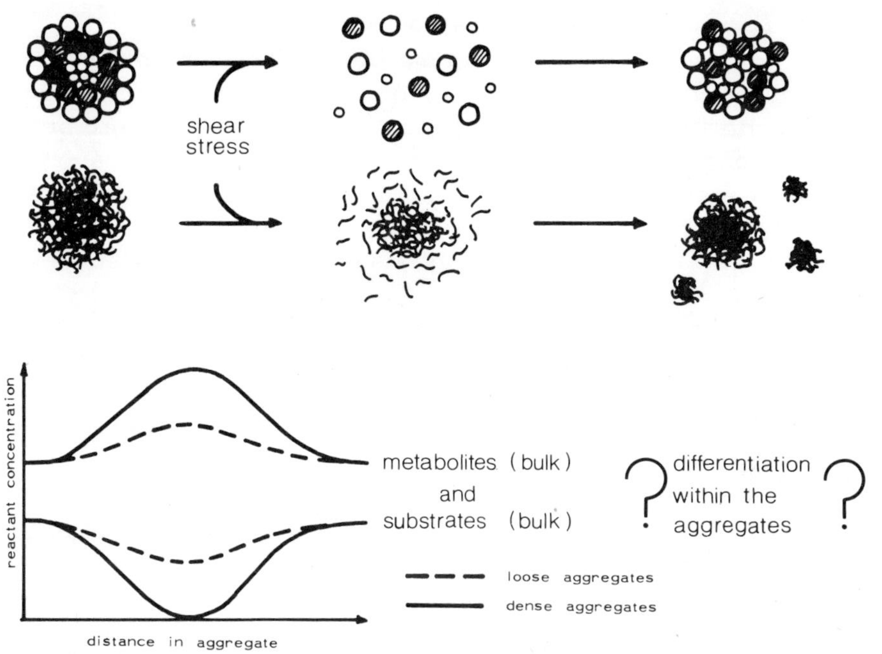

Fig. 3. Problems resulting from aggregation.

transported into turbulent regions of the fermenter.

The filamentous type of aggregation is much more critical with respect to cell ageing and biochemical differentiation. These aggregates are very stable and cannot be easily disaggregated by shear stresses in the fermenter. A clear and definite age distribution through the pellet must exist, with older cells in the centre and the younger fraction on the periphery. In addition, with filamentous organisms growth can only occur on the periphery. However, cells in the centre still have their own active metabolism so that steep concentration gradients of reactants must exist throughout the aggregates. Due to the varied microenvironments faced by the cells, the differentiation of the physiological state of the culture within a pellet is even more pronounced than that comparably found in the different phases of a typical batch culture. In cases where differentiation becomes irreversible, continuous culture of these cells is impossible.

For many fermentations producing secondary metabolites this type of differentiation is perhaps desired, but for many others it may not. From the fermentation point of view, however, no satisfying method of disaggregation has yet been found. One possibility of reducing differentiation phenomena in moulds is to produce loosely flocculated pellets, or pulpy filaments. This condition is perhaps favoured in airlift fermenters. On the other hand, such cultures can exhibit very high apparent viscosities even at low biomass concentrations, since the long filaments greatly affect the viscosity.

Experiments trying to reduce the length of hyphae by imposing high shear stresses on the cells have not usually been successful (Kossen & Metz, 1976). Apparently, high shear stresses tend instead to change the permeability control of the cell membrane (Tanaka et al., 1975; Tanaka & Ueda, 1975). The cells are "ground down" rather than "cut to pieces". New unit operations of cutting filaments should be aimed at in the fermentation process, in order to overcome these problems.

Newly published results from experiments investigating the growth activities in a Streptomyces broth (Fewkes & Wang, 1976) indicate clearly, that the bulk flow to shear ratio is the guiding operating parameter in this type of fermentation. For several reasons it should be much easier to get shear rates and bulk flow under control in the recycle reactor than in a standard configuration fermenter (Fig. 1).

RECYCLE FERMENTER

Now let me go back to the initial point, the fermenter configuration, in particular the recycle fermenter. All recycle fermenters have one common feature: they all try to get control over the mixing in the entire fermenter by control of the bulk flow regime. On the other hand the recycle fermenters shown in Fig. 1 are all different in their realization of aeration and in the pattern of power dissipation. All these features, together, will decide the practical reactor performance and the behaviour of the reactor in scale up.

The conversion in a fermenter will naturally depend on the kinetics of the biological reaction (which is mostly described by a Monod type rate equation), by the distribution of the concentration of reactants, and by their residence time in the reactor. This latter characteristics of the reactor will depend on its bulk flow regime. Bulk flow regimes of biological reactors have been analyzed on a theoretical basis by many investigators (Fredrickson et al., 1970). Two idealized regimes of flow can be classified in order to describe the performance characteristic: the perfectly mixed continuous fermenter and the continuous plug flow reactor (Fig. 4). In the perfectly mixed fermenter all reactants fed into the reactor are immediately mixed and uniformly distributed, so that the total reactor is homogeneous.

The bulk flow characteristics of this fermenter are easily described by the residence time distribution (or the exit age distribution) of reactants passing through the reactor. It describes the variance of the ages or residence times during the flow through the fermenter. It is apparent, from Fig. 4, that in the perfectly mixed fermenter single reactants do not have a defined residence time.

In the plug flow reactor no mixing takes place in the axial direction. Particles entering the fermenter have a defined residence time, and the concentration of reactants is different in its spatial distribution within the reactor. This different mixing behaviour must have consequences for the reactor performance. These consequences have been analyzed by Tsai et al. (1969) (and by many others) for a Monod relationship of reaction rate and a given initial cell concentration in the fermenter feed. As shown on Fig. 4 the substrate utilization and the formation of biomass are affected in a different way in the two reactor types. One can conclude that the perfectly mixed fermenter in the region of short residence times gives rise to higher cell densities and lower residual substrate concentration than the plug flow reactor.

With prolonged residence times, however, the situation is reversed, the plug flow reactor has the highest concentration of microorganisms and the better substrate utilization.

Accordingly it has also been proved by many investigators that for most fermentations a combination of both reactors in a 2-stage process would provide the optimum overall performance (Topiwala, 1974).

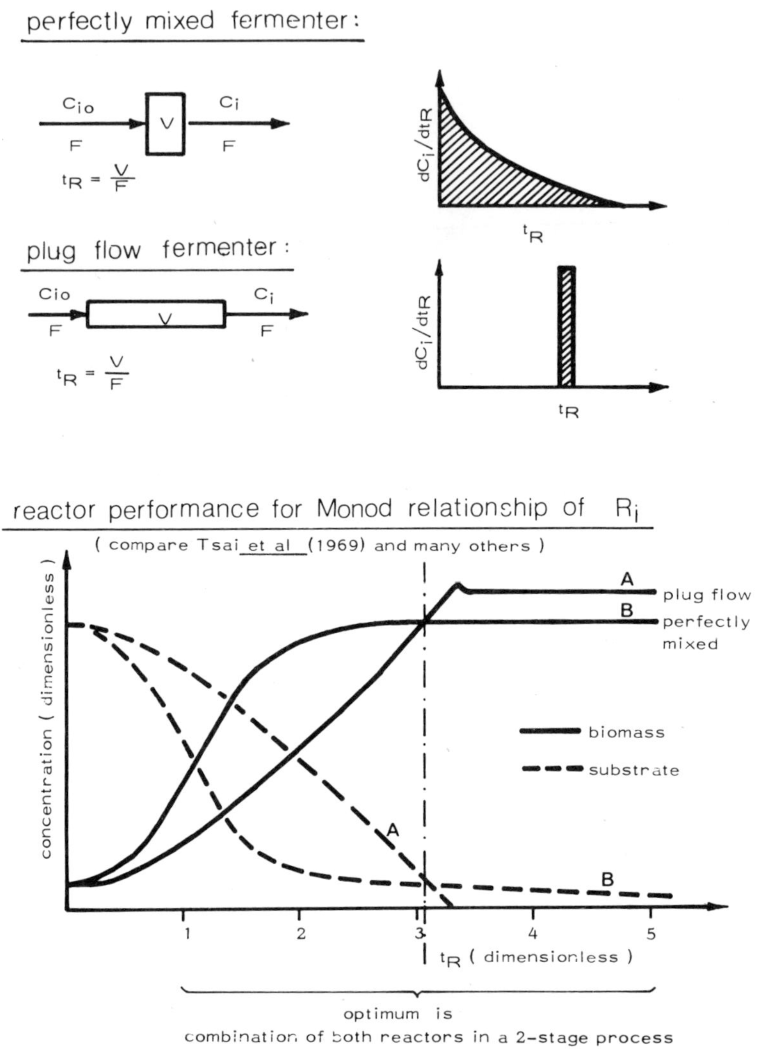

Fig. 4. Idealized flow regimes of fermenters and their performance.

C_{io}	reactant concentration in feed stream
C_i	reactant concentration emerging from reactor
F	volumetric flow rate of feed
t_2	residence time of reactants in the reactor
V	reactor volume

A practical realization of such a desired combination is given in a recycle fermenter. This has been demonstrated by the theoretical analysis of recycle reactors by Rippin (1967), which clearly shows that a recycle reactor is a degenerate case of maximum mixedness and plug flow (Fig. 5).

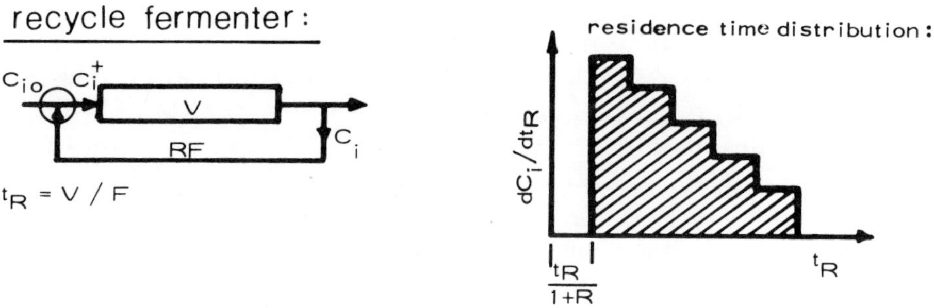

$V/(1+R)F = t_R/(1+R)$ one single passage

compare U. T. Rippin (1967) and many others

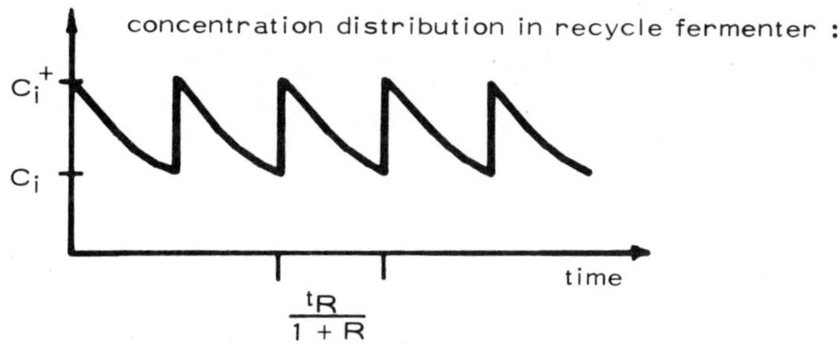

Fig. 5. Flow regime of recycle fermenter.

C_i^+ reactant concentration after feed injection at mixing point of recycle fermenter

R ratio of recycle flow to feed flow

Such a fermenter configuration is realized in practice when a certain fraction of the exit stream leaving a plug flow reactor is recycled back and is mixed with the feed stream entering the reactor. A recycle to flow ratio of $R >$ 10 will guarantee sufficient mixing behaviour in the fermenter (Moser & Steiner, 1974). Such a recycle ratio is easily achieved and in practice is much higher. Depending on the recycle ratio a minimum spatial concentration profile is observed in the reactor. This concentration profile is successively folded over upon itself. The time of a single passage through the fermenter is $\frac{t_R}{1+R}$, and this profile will also characterize the recycle reactor.

The recycle to feed ratio is therefore the first parameter which has to (and can easily) be kept constant in the scale up of a fermentation in a recycle fermenter.

The second main task of the bioreactor which has to be realized in practice, and kept under reasonable control, in a scale up procedure is to provide sufficient gas/liquid mass transfer rates. In a fermentation fluid of given rheologic properties the mass transfer, in particular gas/liquid transfer, correlates with the gas/liquid interfaces, with the fluid velocities at these interfaces, and with their holdup time in the reactor. The creation of the interfaces is dependent on the density of spatial energy dissipation (Hirner, 1974) and is also governed by the manner by which the energy is dissipated into the broth.

As already mentioned repeatedly, all these factors are coordinated and transmitted by the reactor configuration, and it lies mainly in the originality of the designer to fulfill these requirements in the most efficient, transparent, flexible and simple form. One among many interesting and novel fermenter configurations, which fits well to these requisites is the "Waagner-Biro Submerged Reactor" (Fig. 6) (German patent 2627231).

Basically this reactor is a bubble column in which the downcoming broth is aerated during its passage through a self-priming aerator. After the aerator the liquid bulk passes a diffusor in which the self-priming pressure drop is generated. It then sinks down with the given liquid velocity. In the downcomer tube the liquid is required to have a certain minimum sinking velocity, which is dictated by the ascending velocity which the bubbles would have in a stationary broth. The resulting sinking velocity of the bubbles is therefore much lower than that of the liquid. When the cross sectional area of the riser tube is kept smaller than that of the downcomer the ascending velocities of both bubbles and broth are high enough to give reasonable drive in the degassing cyclone, from which the degassed broth is recycled

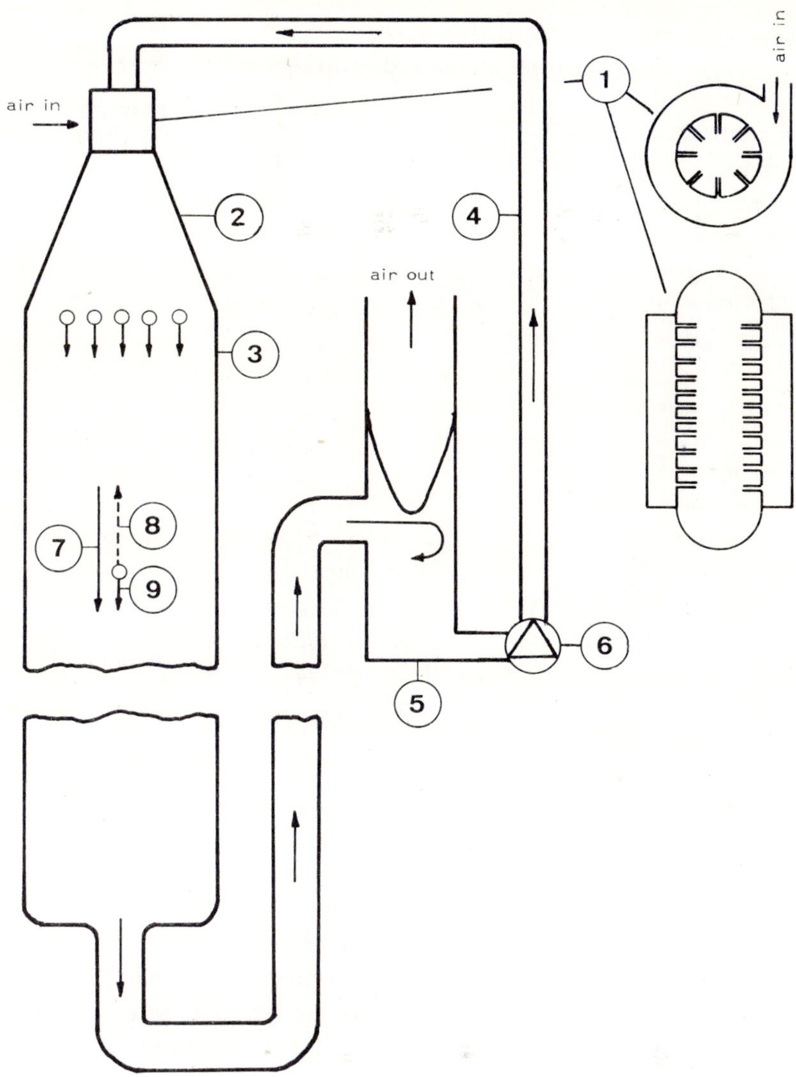

Fig. 6. The "Waagner-Biro Submerged Reactor" - Scheme of flow and function.
1 - aerator made of elastic material; 2 - diffusor; 3 - downcomer pipeline; 4 - riser pipeline; 5 - cyclone; 6 - centrifugal pump; 7 - liquid velocity; 8 - ascending velocity of gas bubbles in stationary broth; 9 - resulting sinking velocity of gas bubbles.

through the aerator by a pump.

The system has a considerable selfbalancing capacity. The minimum necessary sinking velocity of Candida yeast broth in the downcomer pipeline, which will still result in fully dispersed gas/liquid flow, is approximately 25 cm/sec (Katinger, 1973). Under these conditions more than 50 % of gas hold-up can be achieved in the downcomer, and a nearly infinite hold-up time can be achieved. Another opportunity to influence the gas/liquid interface is given, within certain limits, by the mixing ratio of air to broth in the aerator on the one hand, and also by the energy dissipation density within the aerator on the other.

This latter possibility is of great advantage for fermentation processes which require high gas/liquid mass transfer capacities. In those cases the reactor is equipped with an aerator made of elastic material that allows a continuous variation of the cross sectional area similar to a pinch-cock (Fig. 6). When the cross section is small, at a given throughput of broth, the energy dissipation density in the aerator will increase and produce a spectrum of small bubbles, which results in an increased gas/liquid interface, shortened gas hold-up and increased oxygen transfer. The opposite behaviour is also possible, for fermentation mashes characterized by low bubble coalescence rate and bad degassing properties. By lowering the energy dissipation density at constant bulk flow through the aerator, a spectrum of bigger bubbles, with improved degassing speed, is achieved.

At this point I must however stop enumerating the various advantages offered by this configurational conception, since it has never been my intention to favour one single fermenter configuration in my lecture. This configuration has simply been introduced as an exemplary case of modern fermenter design, in which really all of the functional and economic features required for successful fermenter design have been realized. The functional and economic flexibility of this recycle fermenter configuration is mainly based on the fact that the several unit operations necessary in a fermenter, such as aeration and creation of interfaces (gaseous, liquid and solid), energy distribution-density input, bulk flow etc., are separated from each other and not concentrated into one multifunctional unit as in the case of the conventional stirred tank reactor.

The main reasons that recycle type fermenters appeal to me are: they facilitate both the realization of the overall objectives and the scale up of the fermentation.

REFERENCES

Cooney, C. L., Wang, D. I. C. & Mateles, R. I. (1969). Measurement of heat evolution and correlation with oxygen consumption during microbial growth. Biotechn. Bioeng. 11, 269-281.

Fewkes, R. C. J. & Wang, D. I. C. (1976). Physical parameters governing the behaviour of non-newtonian mycelial antibiotic fermentations. Fifth Intern. Ferment. Symp. Berlin, Dellweg, H. (ed.), 77, Berlin, Verlag Versuchs- und Lehranstalt für Spiritusfabrikation und Fermentationstechnologie, Institut für Gärungsgewerbe und Biotechnologie.

Fredrickson, A. G., Megee, R. D. & Tsuchiya, H. M. (1970). Mathematical models for fermentation processes. Perlman, D. (ed.), 399-485, Advances in Appl. Microbiol. 13, Academic Press.

Hirner, W. (1974). Stoffübergang und Stoffübergangsfläche in Gas/Flüssigkeitsreaktoren. Ph. D. Thesis, Univ. Stuttgart.

Imperial Chemical Industries (1971). Fermentierungsverfahren sowie Vorrichtung zu dessen Durchführung. German Patent, Offenlegungsschrift 2135762.

Katinger, H. W.-D. (1973). Simulierung, Differenzierung und Bestimmung physikalischer Fermentationsparameter in Rezirkulationssystemen. 3. Sympos. Techn. Mikrobiol., Dellweg, H. (ed.), 95 - 100, Berlin, Verlag Versuchs- und Lehranstalt für Spiritusfabrikation und Fermentationstechnologie, Institut für Gärungsgewerbe und Biotechnologie.

Karrer, D., Einsele, A. & Fiechter, A. (1976). Reactor design for high mass transfer rates and mixing efficiency. Fifth Intern. Ferment. Symp. Berlin, Dellweg, H. (ed.), 67, Berlin, Verlag Versuchs- und

Lehranstalt für Spiritusfabrikation und Fermentationstechnologie, Institut für Gärungsgewerbe und Biotechnologie.

Kossen, N. W. F. & Metz, B. (1976). The influence of shear upon the morphology of moulds. Fifth Intern. Ferment. Symp. Berlin, Dellweg, H. (ed.), 78, Berlin, Verlag Versuchs- und Lehranstalt für Spiritusfabrikation und Fermentationstechnologie, Institut für Gärungsgewerbe und Biotechnologie.

Moser, A. & Steiner, W. (1974). Verweilzeitverteilungsverhalten eines Rohrreaktors mit Rückführung. Chem. Ing. Techn. 46, 695.

Rippin, D. W. T. (1967). The recycle reactor as a model of incomplete mixing. Ind. & Engin. Chem. Fundamentals 6, 488-492.

Schreier, K. (1974). Bio reactors: stage of development and industrial application, especially with regard to systems for transfer of gas. Proceedings of the Fourth Intern. Symposium on Yeasts, Part I, Vienna, Austria.

Shu-Chien (1975). The red blood cell. 2nd Edn., Vol. 2: Aggregation of blood cells in suspensions. Academic Press.

Smidt, Ena (1976). Untersuchungen über die Hefeaggregation und ihre Einflußfaktoren. Diploma Thesis, Univ. of Agriculture, Vienna, Austria.

Tanaka, H., Takahashi, J. & Ueda, K. (1975). A standard for the intensity of agitation shock on mycelia on agitation of mycelial suspensions. J. Ferment. Technol. (Japan), 53, 18-26.

Tanaka, H. & Ueda, K. (1975). Kinetics of mycelial growth accompanied by leakage of intracellular nucleotides caused by agitation. J. Ferment. Technol. (Japan), 53, 27.

Topiwala, H. H. (1974). The application of kinetics to biological reactor design. In: Advances in Microbial Engineering. Biotechnol. Bioeng. Symp. Suppl. 4 (2), 681-690.

Tsai, B. I., Erickson, L. E. & Fan, L. T. (1969). The effect of micromixing on growth processes. Biotechn.

Bioeng. 11, 181-205.
Waagner-Biro-AG (1976). Verfahren und Einrichtung zur Behandlung einer Flüssigkeit. German Patent, Offenlegungsschrift 2627231.

PRODUCTION OF INDUSTRIAL ENZYMES

K. Aunstrup

INTRODUCTION

Microbial enzymes are probably the most popular topic of research today both in basic and applied biochemical and microbiological science. As a result a vast amount of knowledge has accumulated with regard to the properties of the enzymes, the mechanism and regulation of their synthesis and the genetics involved in their formation.

The impact of this development on the enzyme industry may seem surprisingly small when viewed from the outside. From the point of view of the industrial microbiologist the situation looks different. He has learned that the transfer of scientific discoveries to industrial applications is limited by economic and technological possibilities. Furthermore, he will often experience that the elegant methods described in the scientific literature are not readily applied to the complex systems he is forced to use.

In the following some of the problems facing the enzyme industry today and the forces which govern the evolution of the industry will be described.

ECONOMY

Industrial enzymes have been given a lot of publicity lately, good as well as bad, but very little and often erroneous information has been given about the economy of the industry.

An estimate of the world sale of microbial enzymes from the beginning in 1894 when mold bran or Takadiastase was introduced in the US is shown in Fig. 1.

The first part of the diagram is, of course, highly inaccurate but serves to show that sales were small and relatively unimportant from an economic point of view up till about

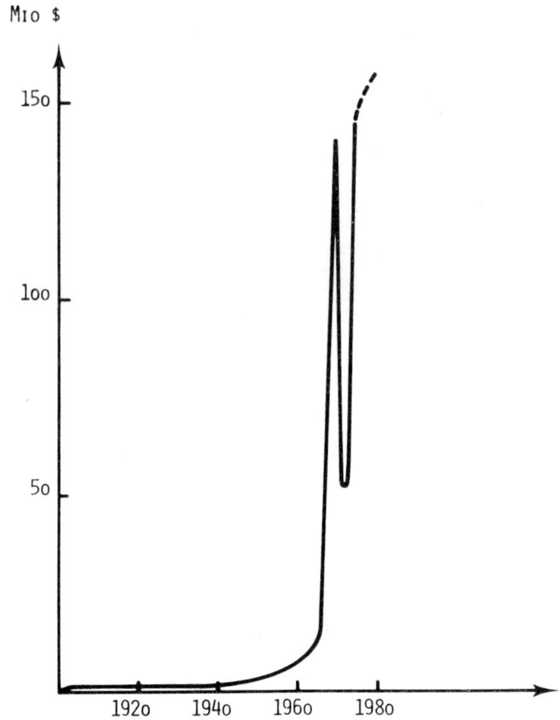

Fig. 1. Microbial enzyme sales world-wide.

1965. At that time the detergent enzymes came into general use. This created an explosion in the industry. Everybody wanted to make enzymes, and everything could be sold. But in 1971 the backlash came. Allergic symptoms were discovered in some workers handling enzymes in a detergent factory.

The public was scared, so enzymes were taken out of most detergents, particularly in the U.S.A., and enzyme sales dropped dramatically. The risks in handling enzymes were then checked very carefully (NAS 1971), and it was found that with proper precautions in the factories and by encapsulating the enzymes before they reach the consumer, every risk can be eliminated. Now the application in detergents has reached a sound level, and there is a steady growth in the use of enzymes in that part of the detergent market where enzymes can improve washing results.

In the meantime the starch industry has become an important outlet for enzymes. Alpha-amylase and amyloglucosidase have substituted acid completely in the manufacture of dextrose, and during the last few years we have seen a really exciting development - the introduction of glucose isomerase, which makes fructose from glucose, and which is used in an immobilized form. (Barfoed, 1976).

This development and other new applications such as microbial rennet means that at present enzyme sales are better than the maximum in 1970-71, about 150 mio. $ per year. This is a surprisingly small figure when one considers the value of the products made by the enzymes.

Fig. 2 gives an impression of the physical size of the production of individual enzymes. The world production per year is expressed in tonnes of pure, active enzyme protein.

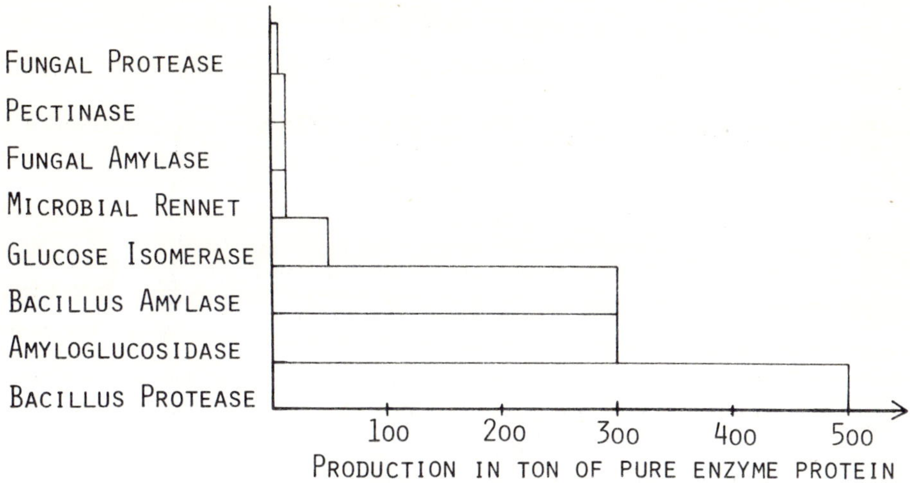

Fig. 2. World production of individual enzymes.

Bacillus protease is dominating, followed by amyloglucosidase and alpha-amylase, whereas the rest of the enzymes are made in comparatively small amounts. This diagram does not reflect the value of the products. Because glucose isomerase is used in immobilized form its value is many times that of a single-use enzyme. Also for the traditional enzymes there is no direct correlation between amount of pure enzyme protein and price. The range of industrial enzymes produced by Novo, and the organisms used in their production, are shown in Table 1.

People often ask why only such a small number of enzymes are of industrial importance. The answer is quite simple.

TABLE 1

Industrial Enzymes Produced By Novo

Enzyme	Source	Trade Name
alpha-amylase	Bacillus subtilis	
	Bacillus licheniformis	Termamyl
	Aspergillus oryzae	Fungamyl
Amyloglucosidase	Aspergillus niger	
Glucose Isomerase	atypical B. coagulans	Sweetzyme
ß-glucanase	Bacillus subtilis	Cereflo
	Aspergillus niger	
Cellulase	Trichoderma viride	
alpha-1,3-glucanase	Trichoderma harzianum	Mutanase
Dextranase	Penicillium sp.	
Hemicellulase	Bacillus subtilis	
	Aspergillus niger	
Lactase	Saccharomyces fragilis	
Pectinase	Aspergillus niger	Pectinex
Protease	Bacillus subtilis	Neutrase
	Bacillus licheniformis	Alcalase
	alkalophilic Bacillus sp.	Esperase
		Savinase
Rennet	Mucor miehei	Rennilase
Uricase	Bacillus fastidiosus	

First of all there must be a demand for a new enzyme, and somebody must have an economic advantage in using the enzyme. Then the technology must be available. The development of glucose isomerase could not have taken place had it not been a fact that the difference between sugar prices and starch prices was sufficiently large, and the technology of immobilized enzymes was available due to the scientific developments. At the moment several interesting possibilities await technological development; for example, the use of enzyme processes in which coenzymes, like NADH, are involved. Hopefully these processes will soon be within reach.

DEVELOPMENT

The enzyme manufacturer must constantly develop new products in order to expand business and to keep abreast of the competitors. I shall not go into details about the work we do to select microorganisms which produce a certain enzyme. I only want to mention that we think a good direct and efficient analytical method is extremely valuable. It should be as close to the application method as possible. The little gadget shown in Fig. 3 was made to make quick tests of rennet enzymes. The holes are connected in pairs through a narrow channel. By tilting the plate, milk will run from one hole to the other until it coagulates.

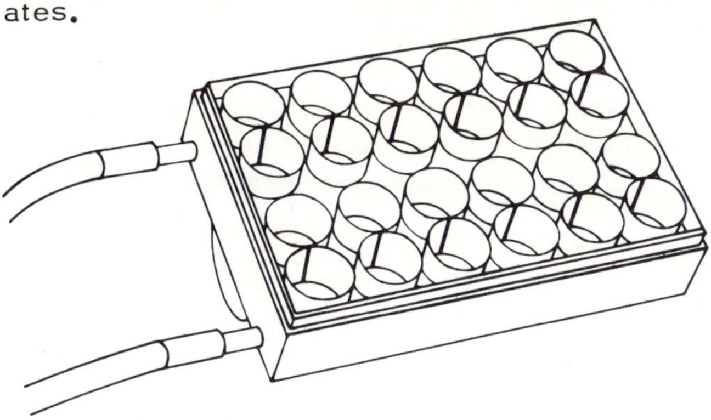

Fig. 3. Instrument for determining milk coagulating activity for screening purposes.

Substrate: 12 % skim milk powder in 0.01 M $CaCl_2$ preheated at 30°C.

Method: Temperature is maintained at 30°C by circulating water through the lower part of the instrument. Enzyme solutions are added to the holes (between 10 and 100 µl). 5 ml milk is added at intervals of e.g. 6 sec. The instrument is rocked at a constant speed, and the time of coagulation is observed. Enzyme activity is proportional to the reciprocal coagulation time. An enzyme standard is used to convert the activity to standard units. For additional details see text.

In this way, a large number of samples can be tested simultaneously for milk coagulating activity.

When we have found a new organism which produces an interesting enzyme it must be classified. This is important both for patent reasons and also to enable us to evaluate the potential risk involved in handling the organism. Unfortunately, our isolates do not fit well into the known taxa. There is often, if I may say so, too big a distance between the old culture collection strains and the strains of real life. Therefore we, too, need more specific classification systems.

Finally, before we cultivate the organism on a large scale, we make preliminary toxicity and pathogenicity studies on animals to protect our personnel from unnecessary risks.

At this point we can start chemical examinations of the enzyme and application studies in the laboratory. If the results are positive, strain development can begin. The purpose of this work is listed in Table 2. Apart from the obvious yield improvement, we are very much concerned about the removal of undesired side products, such as odour, colour and metabolites like acids or other enzyme activities, which are unnecessary or harmful in the application of the enzyme.

TABLE 2

Objectives of Strain Development
<u>Optimization of desired properties</u>
high enzyme yield
genetic stability
independence of inducer
better utilization of medium compounds
good recovery
<u>Removal or suppression of undesired properties</u>
harmful metabolites
antibiotics
odor, color and slime
unnecessary enzyme activities

I can give you an example of this: Amyloglucosidase is

used in the manufacture of dextrose. Most fungi also make a transferase, transglucosidase, which will catalyze the formation of isomaltose. This will reduce the dextrose yield. Therefore, the amyloglucosidase must not contain transglucosidase. We designed a simple method for the selection of transglucosidase-free mutants (Fig. 4).

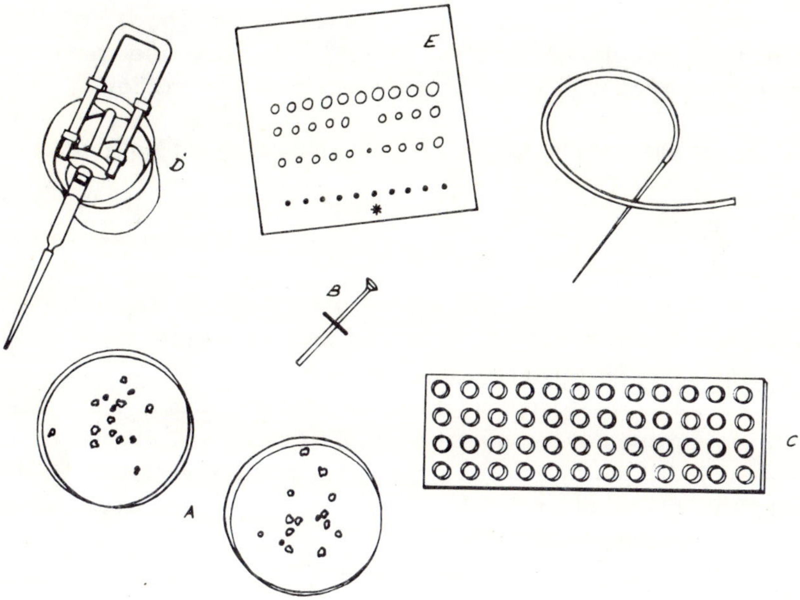

Fig. 4. Screening for transglucosidase-free mutants. See text for details.

Spores, mutagenized with nitrosoguanidine or UV-light, are spread on agar (A). Plugs are taken from each colony by the small cork borer (B) and transferred to the plexiglass plate (C). A small amount of maltose (D) is added, and the plate is incubated for two hours. The amyloglucosidase in the colonies and the agar will now react with the maltose and form glucose, whereas the transglucosidase will form isomaltose. The reaction mixture is transferred to the thin layer chromatography plate (E), and after proper drying, development and staining we see the result: All the colonies, except for one marked X, form transglucosidase. This is our mutant, and we can go back to the plexiglass plate and take it out for recultivation and purification. This may seem to be a very laborious method but the fact is that one person can test a thousand colonies a day, when the work is properly organized.

If you ask me whether we use more sophisticated genetic methods in our work, I am afraid I have to disappoint you. We have made attempts, and we are following developments within the area, but so far we have found that straight forward mutagenization, combined with good selection method, is the fastest way to reach our goals. The primary excuse for this is, of course, that our strains are genetically unknown.

DEVELOPMENT OF MEDIUM AND FERMENTATION METHOD

The next step of the development work is to find a suitable combination of fermentation medium and fermentation conditions.

In the literature one often finds papers describing medium optimization made in shake flasks or laboratory fermenters, and complicated media with many and often expensive components. Such investigations may be valuable for the intended purpose, but they are of little use in industrial fermentations.

The raw materials for industrial fermentations are limited to substances which are readily available in large quantities at a low cost, and what is very important: they must be available in food or feed quality. This limits the number of raw materials to the ones listed in Table 3. In addition it will, of course, in many cases be necessary to use various inorganic salts.

TABLE 3

Suitable Medium Constituents for Industrial Fermentations
Starch hydrolysate, saccharose, molasses,
corn, wheat, barley,
soybean, cotton seed meal, peanut meal,
corn steep liquor, whey.

Typical industrial media will often be simple in composition like the one for microbial rennet, given in Table 4, but they will often be characterized in having a high concentration of dry substance. The reason is, of course, that there is usually a correlation between dry matter content, cell growth and enzyme yield.

TABLE 4

Medium for Production of _Mucor miehei_ Rennet

Patato starch	4 %
Soy bean meal	3 %
Ground barley	10 %
$CaCO_3$	0, 5 %

In many cases it is an advantage to start with lower concentrations and feed medium constituents during the fermentation.

In fermentations where glucose represses the enzyme formation, such as is the case with alpha-amylase, it is imperative to adjust the carbohydrate addition to the medium so that no glucose builds up. This is apparently not the case for amyloglucosidase, as seen in the fermentation diagram of Fig. 5. A high glucose concentration builds up in the medium shortly after inoculation, due to hydrolysis of the starch in the medium, but enzyme production continues unaffected and comes to an end shortly after the glucose has been used up.

It is interesting to notice the oxygen level, which drops to almost zero, when enzyme production starts, and remains at a low level during the whole enzyme production phase. This may indicate that oxygen transfer (or in other words, fermenter performance) is limiting the production. In many cases this is the limit for the industrial microbiologist. He knows the capacity of the production plant and has to design his medium and fermentation methods accordingly.

When should a fermentation be stopped? As you will see, the enzyme activity curve goes straight up to a deflection point, then flattens. Had the fermentation run longer, we would have seen a fall in enzyme activity in the end.

The most logical point to stop at would be the deflection point, but in many cases one will obtain a better recovery

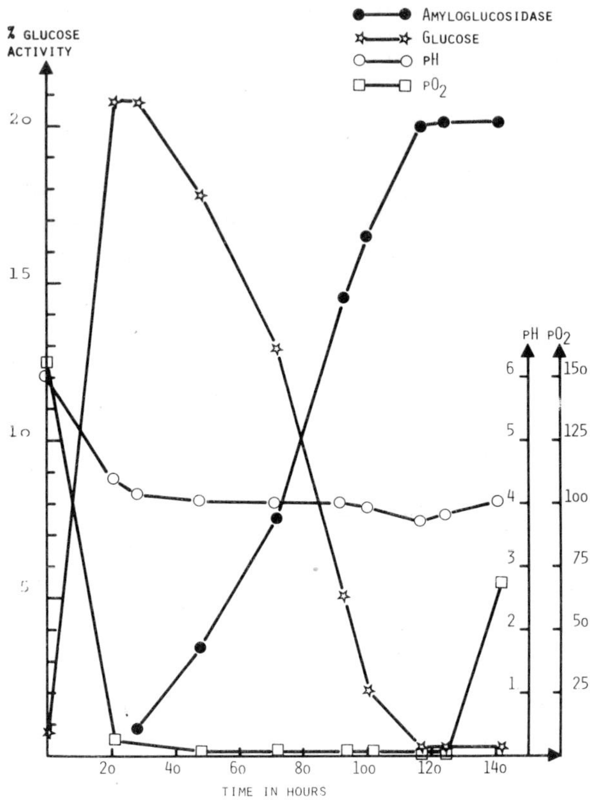

Fig. 5. Time course of amyloglucosidase formation.

process, if the fermentation is taken a little further, so that more substances in the medium are metabolized.

Straight forward batch processes like this are traditionally used for enzyme production. The process may be improved by feeding one or more components during the fermentation. Pure continuous processes are rarely used in large scale industrial enzyme production. The only reported example in the literature is Novo's production of glucose isomerase (Diers, 1976).

DEVELOPMENT OF A RECOVERY METHOD

At the end of the fermentation process, our ideal is a liquid with a high enzyme concentration, mycelium which is easy to separate, and as little other soluble products as possible. In most cases the broth will contain 0.5–1 % useful enzyme protein, 1–1.5 % mycelium and 5–10 % soluble medium components. The methods used to develop a suitable recovery process will depend on the enzyme and its uses, and it will be limited by the equipment in the factory or the possibilities for investment in new equipment.

The requirements for the finished product can be grouped under three headings: Safety, Quality and Reliability.

A safe enzyme product should have low allergenic potential. This means that the product should be liquid or appear in dustfree form. Furthermore, it should be free from toxic materials and harmful microorganisms. A high quality enzyme product should be easy to handle and use. The enzyme activity should be standardized and odour, colour and other undesired properties should be absent. Reliability means that the enzyme activity is stable during storage and that the preparation is resistant to microbial attack.

There is usually no requirement for high purity of the enzyme. For most applications this is neither important nor desirable, because high purification is usually accompanied by reduced stability under storage and application.

When the development work is over, a careful toxicological testing must be made on a mixture of several batches, preferably of production size. Some of the tests which have to be performed are shown in Table 5.

Finally appropriate specifications are set up to ensure that the product is of uniform quality. Last but not least, a price must be fixed.

PRODUCTION

Most enzymes are produced in submerged culture in traditional fermenters. The important factors in today's fermentation technology are: automatic control of the fermentation parameters; sufficient aeration capacity, and possibility of automatic feeding of medium constituents during the fermentation.

We are still to a large extent dependent on manual sampling and analysis. This results in a time lag, which often creates problems. On-line detection of microbial contamina-

TABLE 5

Biological Tests on New Enzymes

TEST:	SPECIES:
<u>Oral toxicity</u>	
Acute:	Mouse, rat
4 week:	Rat
12 month:	Dog
24 month combined with carcinogenicity and teratogenicity:	2 rodents
<u>Cutaneous toxicity</u>	
Skin and eye irritation:	Rabbit
<u>Pathogenicity of production strain</u>	
Inoculation:	4 animal species
<u>Antigenicity</u>	
Relative antigenicity, sensibilisation:	Guinea pig
Sensibilisation:	HOMO
<u>Clinic</u>	
Health control, personnel:	HOMO

tion, measurement of carbohydrate concentration and enzyme activity would be a great help. In principle the problem could be solved with the equipment available today. The reliability is, however, still not sufficiently high.

In the production of enzymes from filamentous fungi, the mycelium form is important for power consumption, aeration and yield. The filamentous mycelium demands a higher power input to the fermenter than the mycelium in pellet form. The mycelium form can often be selected by mutation. The pellet form is preferred because of its smaller energy requirements but also, because it is easier to remove in the recovery process.

RECOVERY

One of the simplest processes for making an enzyme product from fermentation broth is shown in Fig. 6, where the preparation of a liquid enzyme product is shown.

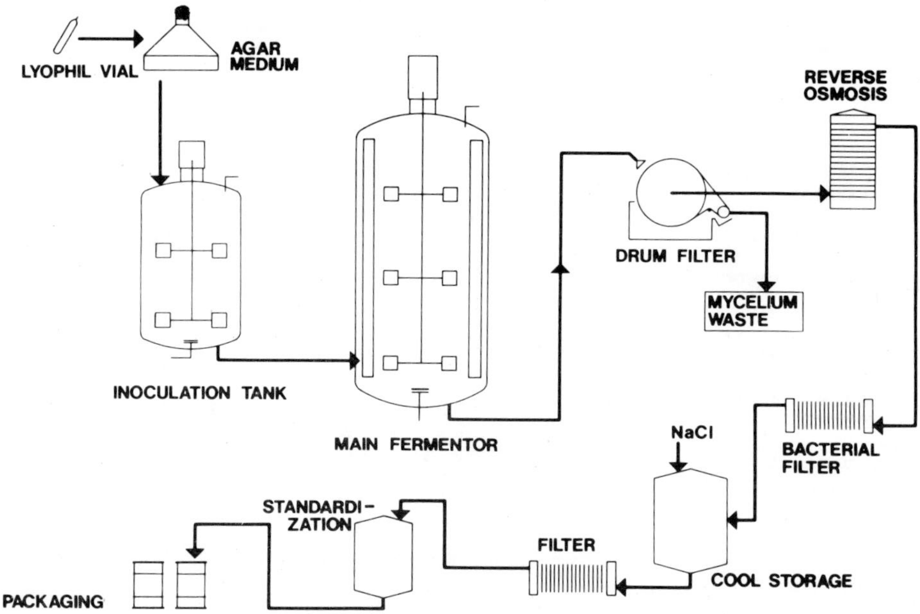

Fig. 6. Production of a liquid enzyme preparation.

The process steps are: removal of mycelium, concentration by reverse osmosis, bacterial filtration, addition of the preservative NaCl, standardization and packaging. Most enzymes are prepared in a liquid form by a simple recovery process like this. It is an advantage to the manufacturer, because the process requires few investments, and is inexpensive to operate. Liquid products are also preferred by the customers, because they are easy to use, and because the allergy risks are minimal, due to the absence of dust.

If a solid product is necessary, such as for example with the detergent enzymes, the recovery process becomes much more complicated, as shown in Fig. 7. A precipitation step is introduced. Salt precipitation is shown in the diagram, but solvent precipitation could equally well have been used.

Thereafter we have a series of operations which serve the purpose of embedding the enzyme to prevent dust formation. This diagram shows a prilling process: the enzyme is mixed with wax and sprayed into a chamber, where small drop-

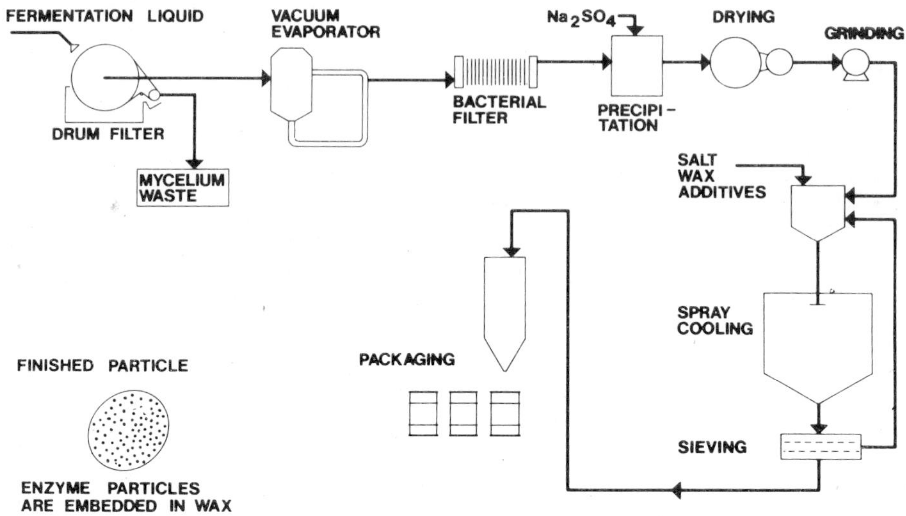

Fig. 7. Production of dustfree enzyme.

lets are formed; they solidify by cooling, and after sieving they are ready for packing. The enzyme is almost evenly distributed in the final particle.

Abrasion during handling may give minute dust formation, and therefore it may sometimes be difficult to meet the strict limits for dust formation set today. The limit is less than 0.05 μg of pure enzyme protein dust per 100 g of enzyme preparation. An improved method has therefore been developed in which the enzyme-containing granule is coated with an exterior layer of wax through a series of operations.

The resulting product has an extremely low dust level. It can be said without going into details that processes like this are difficult to operate. They are very demanding with regard to quality and uniformity of enzyme concentrate, additives and the skills of the personnel operating the plant.

The latest development, the immobilized enzymes, raises new problems in the recovery process. A large number of immobilization methods are known but only a few have been used in large scale. Of these, crosslinking with glutaraldehyde is the most widely used. The method is simple in principle: enzyme-containing cells are, if necessary, homogenized. Then they are mixed with glutaraldehyde and in some cases inert protein to aid the crosslinking reaction. The resulting gel is shaped into suitable granules which are washed and dried.

The physical properties of an immobilized enzyme are very important. In order to permit good column operation, the particles must be of uniform size, they must be strong, and the porosity must be so high that the enzyme can express its activity and maintain it for a long time. Glucose isomerase, for example, is usually used for over 1.000 hours at a temperature between 60 and 65 $^{\circ}$C.

CONCLUSION

It has been exciting to work within the enzyme industry during the phantastic development of the last fifteen years. Partly because of the rapid growth of the industry, but most of all because we have experienced a rapid utilization of scientific results both from molecular biology and enzyme chemistry.

Science gives us the basic knowledge and the new outlooks. The industrial worker makes things possible by selecting those ideas which are realistic with the technology available. He makes the processes economic by an often painstaking work on optimization of strains and fermentation and recovery methods. A close co-operation between basic science and industry is necessary to obtain a good result. Let us hope we can continue the fruitful co-operation of the past also in future.

REFERENCES

Barfoed, H. C. (1976). Enzymes in Starch Processing. Cereal Foods World, 588.

Diers, I. (1976). Glucose Isomerase in Bacillus coagulans in Continuous Culture. Dean, A. C. R., Ellwood, D. C. Evans, C. G. T., Melling, J. (eds.), 6, 208, Ellis Horwood.

NAS (1971). PB 204118, Report of the ad hoc Committee on Enzyme Detergents. Washington.

LIGNOCELLULOSE FERMENTATIONS

C. Szajer and Z. Targoński

INTRODUCTION

In recent years there have been published many papers and excellent reviews about recent research on the cellulases, dealing with the problems of the synthesis as well as with the composition of the cellulase complexes and their action on different kinds of cellulose. When we take into consideration that the year 1950 is the beginning of microbiological as well as biochemical research on the biodegradation of cellulose, we can see how great the progress had been during these 27 years. The bibliography of this period of time would be a really big volume of papers. Today we can say that this is a rather wide range of problems - so I don't intend to give a detailed and fully documented survey of all topics but rather to correlate some previous and recent works.

In 1975 in Aulanko (Finland) the last international symposium on the enzymatic hydrolysis of cellulose was held. Papers presented there had shown that the area of research of cellulolysis is connected with the knowledge of cellulose chemistry in its microbial, enzymological and bioengineering aspects. Moreover, reports presented at that symposium had shown, that the problems of biodegradation of cellulose are not only a research topic with a long history but that cellulases are being increasingly applied for the controlled modification of foodstuffs or the saccharification of waste cellulose. Besides, the meeting in Aulanko made it possible to reach some common agreement on many questions at issue. The most important conclusion, which was reached by many researchers working on cellulase problems, was that today we have enough evidence that the central enzyme in all cellulolytic systems is an *endo*-beta-1, 4-glucanase, usually associated with an *exo*-beta-1, 4-glucanase and beta-glucosidase. Since the hydrolytic enzymes may be described as *endo*-, *exo*-glucanases and beta-glucosidases the former terms "C_1" and "C_x" can be abandoned.

The possible mechanism of cellulose breakdown can be represented as follows:

native cellulose	endo-glucanase --------------→	cellulose modified
cellulose mod.	exo-glucanase --------------→	cellobiose
cellobiose	glucosidase --------------→	glucose

Different amounts of endo-glucanases in different culture filtrates or cellulase preparations can be found. Eriksson & Pettersson (1975a, b) could clearly distinguish five endoglucanases in the culture liquids of *Sporotrichum pulverulentum*. Pettersson (1975) found two endo-glucanases in a crude cellulase from *Trichoderma viride*.

During the meeting in Aulanko, it was suggested by J. Goksøyr, that one can express the enzyme activity as:

1. cellulolytic activity - in culture filtrates or commercial brands in practical units e.g. filter paper units
2. cellulase activity - meaning the activity of isolated cellulase protein which should be measured on chemically defined substrates in accordance with standard rules.

This is an important question especially for microbiological research, where most of the experiments carried out on cellulase synthesis are concerned with such fundamental points as the effect of culture conditions (pH, nitrogen source, substrate concentration, aeration and temperature). In these cases the term cellulolytic activity for the tested culture filtrates has to be used. In that way we will distinguish in a practical manner the microbiological investigations from the biochemical discussions about the composition and the mechanism of action of the cellulase complexes.

ENVIRONMENTAL FACTORS DURING CULTIVATION

Microbiological research into the biodegradation of cellulose is today concerned with the understanding of the mechanism of enzyme synthesis, by understanding its penetration through the cell-wall of the microorganisms, and the problems of environmental conditions affecting the activity of extracellular enzymes. Pollock (1962) and van Hofsten (1975) studying this question had confirmed, that it is really very hard to distinguish differences between the cell bound enzymes and the extra-cellular enzymes. With bacteria, Yoshikawa *et al.* (1974) have shown that a single cellulase component of high molecular weight is initially released which is subsequently modified to form 5 different components of smaller molecular weight.

From the beginning of our studies on the enzymatic breakdown of cellulose, we were convinced that the solution of these questions is an important factor in obtaining good yields of "cellulases" in fermentation processes, without overlooking the problems of the selection of active "cellulolytic" isolates and the elaboration of optimal culture conditions for the selected microorganisms.

I should mention here that Pollock (1961) described the excretion of penicillinase into the medium as an enzymatic phenomenon. Nomura, Hosuda & Youshikava (1958) found that amylase excretion by *Bacillus subtilis* was preceded by lysis of the cell-wall. Similar results were obtained on proteolytic enzymes excreted by higher fungi. It was also shown that the environmental conditions may cause changes in the enzyme activity of cellulases. Lyr & Schanel (1964) recorded an increase in the activity of cellulolytic and hemicellulolytic enzymes after passing neutral gas (nitrogen) through a medium in which *Fomes marginatus* was grown. Eriksson, Pettersson & Westermark (1974) have demonstrated that cell-free concentrated culture solutions of four different fungi (*Sporotrichum pulverulentum*, *Polyporus adustus*, *Myrothecium verrucaria* and *Trichoderma viride*) degrade twice as much cellulose in the presence of oxygen as in the absence of oxygen. The authors have also concluded that the oxidation enzymes are relatively unstable and have not yet been found in commercial "Cellulase" preparations.

The significance of the effect of environmental factors on the enzymatic activity has also been stressed in many phytopathological papers (Hanks & Feldman, 1966; Foster &

Weber, 1969; Dixon & Pegg, 1972).

THE ROLE OF OXYGEN AND REDOX POTENTIAL

In our own researches in this field it was interesting to study how some selected amino acids affected the activity of extracellular cellulases during an 8 days growth period. As appears from Fig. 1 the concentration of cysteine (1.3 to 5.4 x 10^{-3} M) in culture media did not significantly change the Eh potential during the first six days of growth. On fur-

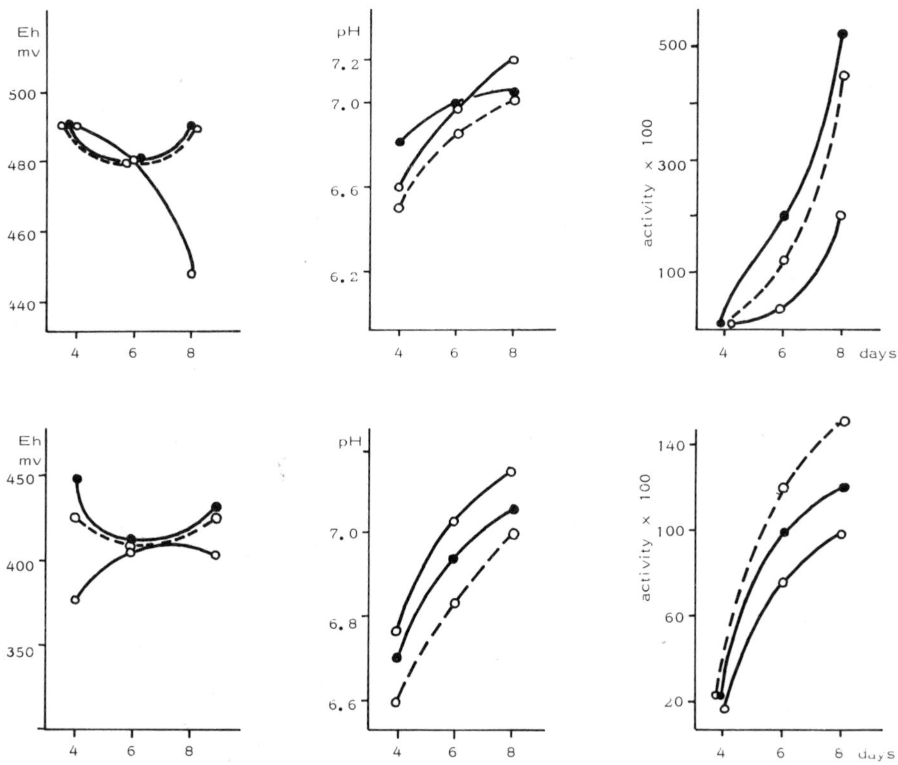

Fig. 1. Effect of cysteine on Eh, pH and enzyme activity in culture filtrates of Fusarium
above: strain 25/6
below: strain 25/2
o-o control, ●-● 1.3 · 10^{-3} M, o--o 5.4 · 10^{-3} M

ther cultivation the media containing cysteine showed higher Eh values than the control without cysteine. At that time also the highest increase in cellulolytic activity in the tested cultures was observed. No significant changes in pH of the media were observed. Similar data were obtained in experiments with a second highly active strain <u>Fusarium</u> 25/2. However the same concentration of cysteine caused smaller differences in the Eh values and simultaneously weaker sti-

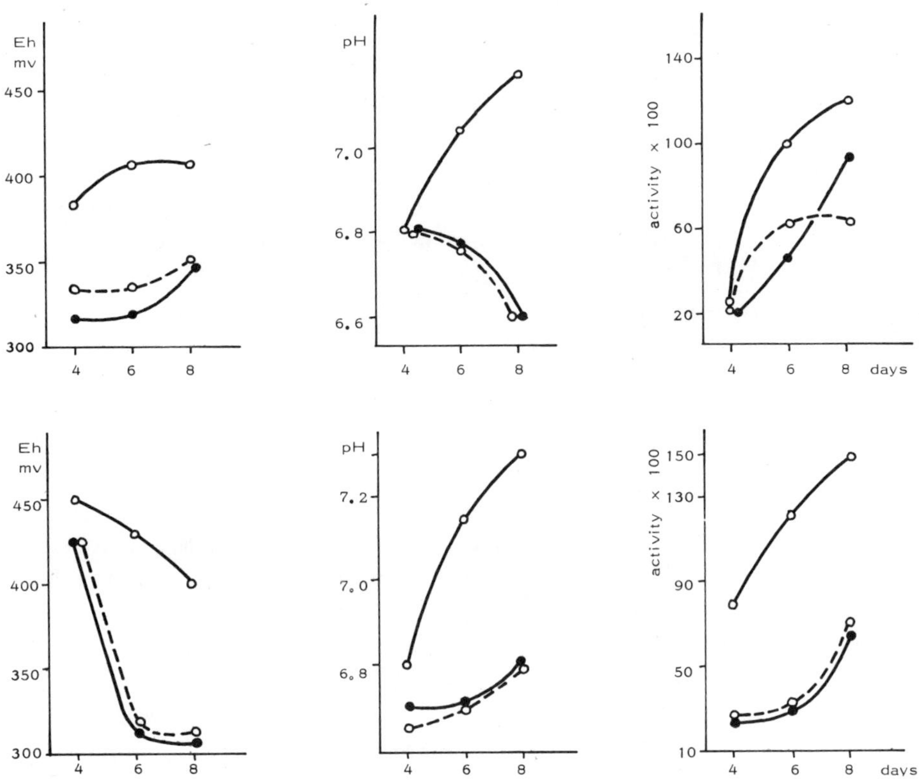

Fig. 2. Effect of thiourea on Eh, pH and enzyme activity in culture filtrates of Fusarium
above: strain 25/6
below: strain 25/2
o-o control, ●-● $1.3 \cdot 10^{-3}$ M, o--o $5.4 \cdot 10^{-3}$M

mulation of cellulolytic activity in this fungus after 8 days of growth (Szajer, 1975). In the experiments with thiourea, a distinct decrease in Eh potential was observed in the culture media of the Fusarium strains 25/2 and 25/6. This was particularly conspicuous between the 4th and 6th day of growth. This phenomenon was correlated with an increase in acidity of the culture media (Fig. 2). However, the changes observed in the pH values were considered too small (pH 7.2 in the control medium and 6.8 in the media with thiourea) to influence the activity of cellulolytic enzymes. We have concluded from the results obtained that the increase in cellulolytic activity in the highly active strains of the fungi studied was connected with a higher Eh potential, and the inhibition of enzyme activity in the presence of thiourea was associated with a decrease in Eh potential.

The changes in the redox potential can also be related with the aeration, i.e. with the concentration of O_2 in the cultivation medium during the growth of the microorganisms. We tested the influence of this parameter on the cellulolytic activity in culture filtrates during a 5 day growth period of the Fusarium 25/2 isolate (Targoński & Szajer, 1977). Testing the effect of aeration it was shown that increasing the amounts of air to the fermentation medium affected the activity of the cellulolytic enzyme complexes. This dependence can be seen when the enzymatic activity of culture media with different aeration levels is compared (Fig. 3). After 5 days of growth of Fusarium sp. in culture liquids which were aerated with 140 ml of air/min the short fibres activity was 3.5 times higher, the beta-glucosidase 2.6 times higher and the endo-glucanase 2 times higher (Fig. 4) than in the culture media aerated with 20 ml of air. It should be mentioned here, that the amounts of cell protein in the culture (after centrifugation), whether aerated with 20 or with 140 ml air, were almost the same. These results suggest that the activity of cellulolytic enzymes increased simultaneously with the growth of Fusarium strain tested and the differences in the cellulolytic activity could be observed only when the degree of aeration was changed. These results will be more intelligible when we take under consideration that the experiments were made using wood-pulp cellulose which con-

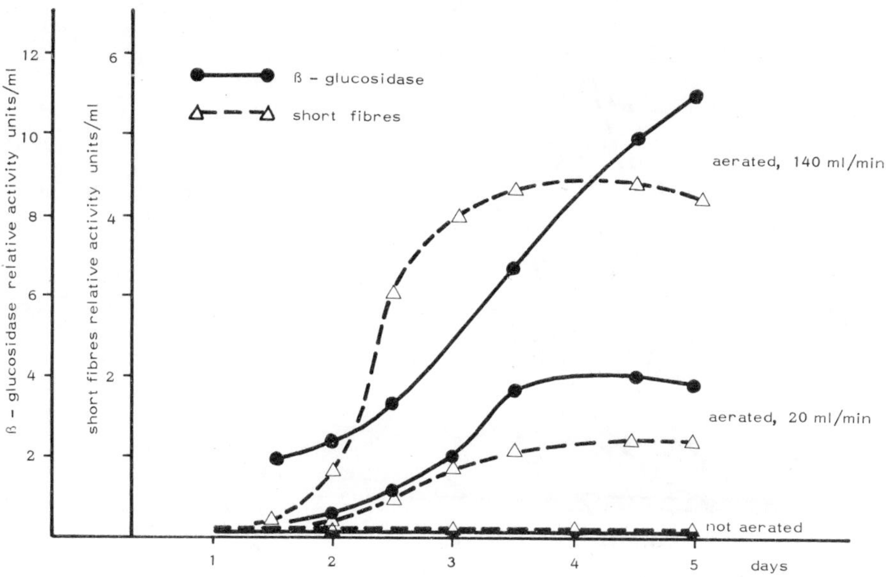

Fig. 3. Short fibres and beta-glucosidase activity in culture filtrates of Fusarium sp. at different aeration levels.

tain about 15 % of lignins, and explain the necessity of oxygen in the culture media not only as a respiration factor but also as a substrate for the cellobiose:quinone oxidoreductase described by Westermark & Eriksson (1974, 1975). This enzyme (obtained from the white-rot Sporotrichum pulverulentum) is an important factor in the oxidation reactions in fungal degradation of cellulose (Eriksson, Pettersson & Westermark, 1974).

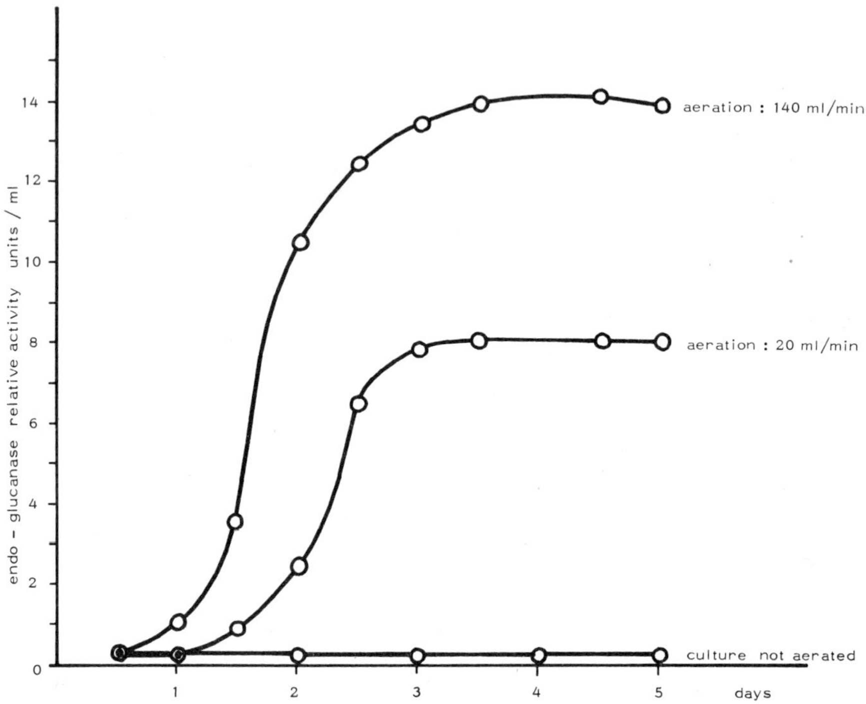

Fig. 4. Endo-glucanase activity/CMC-ase/in culture filtrates of Fusarium sp. at different aeration levels.

INTERRELATIONS BETWEEN CELLULOLYTIC ACTIVITY AND OXIDOREDUCTASES

It was very interesting for us to know whether the Fusarium sp. tested in our laboratory produce enzyme complexes similar to those produced by the Basidiomycetes Sporotrichum pulverulentum. First results dealing with the activity of the oxidizing enzymes in culture filtrates of Fusarium, were obtained by Targoński (1977) and the results are shown in Fig. 5. The data demonstrate that under the conditions used, the fungus was able to synthesize both kinds of enzymes, cellobiose:quinone oxidoreductase and laccase. It was also possible to demonstrate the activity of catalase. Using two different substrates in the cultures - filter paper

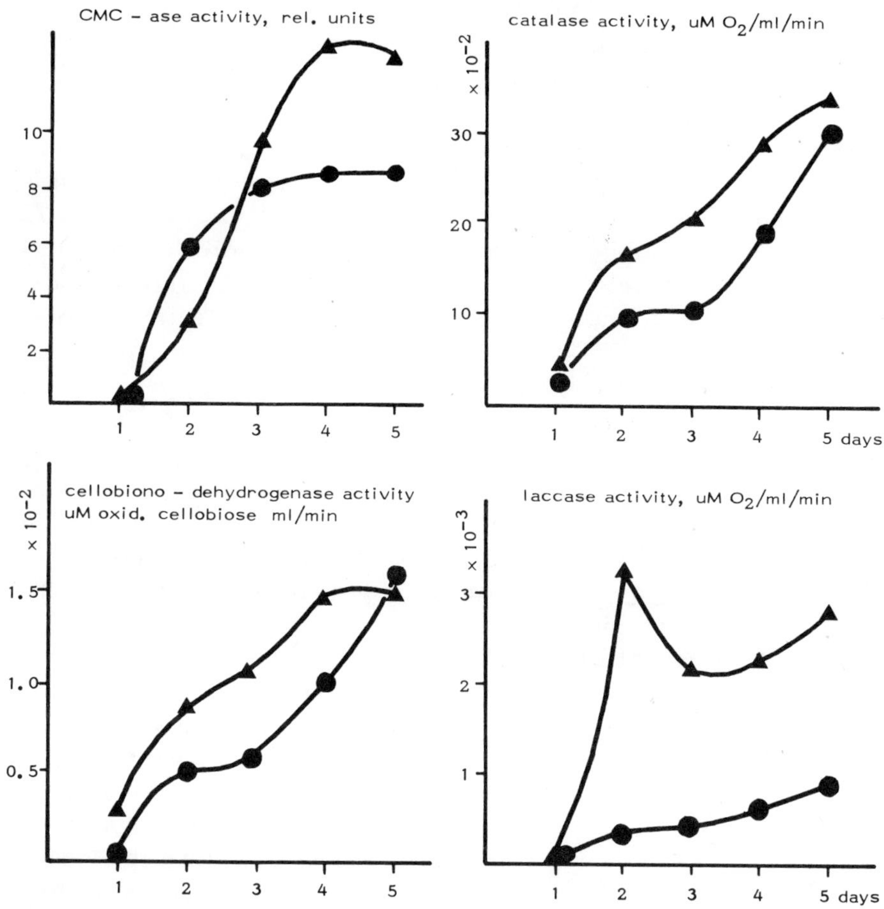

Fig. 5. Enzyme activity of Fusarium sp. grown on 1 % of filter paper cellulose ▲ – ▲ or dewaxed cotton o – o

and dewaxed cotton, in a concentration of 1 % - it could be observed, that the activity of the oxidizing enzymes (catalase and laccase) increased more strongly in the medium containing filter paper cellulose as carbon source. Cellobiono dehydrogenase also shows higher activity in media with filter paper than in the cultures with the dewaxed cotton.

It thus seems very probable - that the same or very

similar mechanisms exist in the Fungi Imperfecti.

Most of the "cellulolytic" microorganisms, for example *Trichoderma viride* (Sternberg, 1976), *Sp. pulverulentum* (Eriksson & Larson, 1975), normally acidified the medium during growth. This is supposed to be the result of the activity of cellobiose:quinone oxidoreductase, causing accumulation of cellobionic acid. In the cultures of our strain the maximum cellulolytic activity occurs in the region of pH 7 - 8. This would suggest that the cellobionic acid is only produced in very small amounts or it is rapidly utilized by the fungus. In the course of my work in Eriksson's laboratory, we found that the cellobionic acid as well as gluconic acid were both utilized very well by *Sp. pulverulentum*. As yet we have not estimated the utilization of cellobionic acid and gluconic acid by the *Fusarium* 25/2 isolate. However these investigations seem to support the idea discussed above (van Hofsten) that some amounts of cellulases may be cell-bound during the whole growth. Our experiments have shown that the *Fusarium* strain studied reached pH 7 - 8 after 3 - 4 days, by which time the cellulose is exhausted, when cultivated on 1 % cellulose. At that time the production of conidia also starts, which is connected with the lysis of the mycelium - because under the microscope no more mycelium can be seen. This is the point at which we always find the maximum cellulolytic activity.

ROLE OF CELLULOSE CONCENTRATION IN THE CULTIVATION MEDIA

In above experiments we regularly used 1 % of wood pulp as the carbon source for the growing fungus. As is well known from numerous papers, increasing amounts of cellulose in culture media always caused a decrease in cellulolytic activity of the culture filtrates (Mandels & Weber, 1969; Ghose *et al.*, 1975; Nystrom & Kornuta, 1975; Eriksson & Larson, 1975; Sternberg, 1976) with no differences when various kinds of cellulose were used. The same phenomenon is apparent from our results (Fig. 6) when different cellulose concentrations were tested (Bujak & Szajer, 1977). As we know, the cellulolytic enzymes are very stable, as can

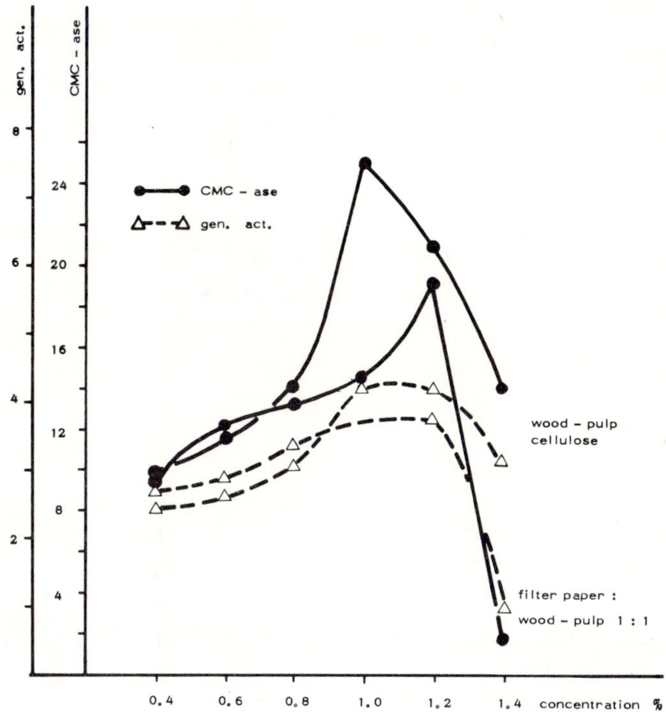

Fig. 6. Effect of different cellulose concentrations on the cellulolytic activity.

be seen from many papers as well as from Table 1, which shows that changes of the activity during a 24 hr period of time at different pH values were very small (Podgórska & Szajer, 1977). Moreover they can be adsorbed on cellulose fibres, from which they can be washed out in an active form.

What is the reason that at cellulose concentrations higher than 0.75 - 1.2 % most of the cellulolytic fungi stopped the synthesis of enzymes, as estimated in culture filtrates which show a decrease in cellulolytic activity. Can we say that the above concentrations are the most effective for cellulases production? The stability of cellulolytic enzymes on the one hand, and the decrease of activity by higher concentration of cellulose on the other hand, interested us a great deal, because it is always important to increase the fermentation ability when ligno-cellulose substrates were

TABLE 1

Stability of Cellulases at Different pH Values During 24 hr

pH	Cellulolytic Activity - ug red. sugars			
	Fusarium		Cellulase-Novo Industri	
	Start	after 24 hr	Start	after 24 hr
2, 5	20	20	100	100
3, 3	50	50	240	250
4, 0	100	90	270	270
4, 4	120	120	280	280
4, 7	150	150	290	290
5, 2	170	165	300	300
6, 0	160	160	280	280
6, 2	140	130	270	270
7, 0	50	50	240	240
8, 5	30	10	190	140

used. We are carrying out some experiments looking for a another explanation of this phenomenon. We assume that the synthesis of the cellulolytic enzymes is continuing at the same level in the course of growth irrespective of the cellulose concentrations. Experiments were carried out with the following cellulose concentrations: 1, 2, 3 and 4 % of cellulose in the growth medium. Three different kinds of cellulolytic activity were examined: a) the endo-glucanase activity - viscosimetrically on CMC, b) aryl-beta-glucosidase using salicin as the substrate, c) the hydrolysis of dewaxed cotton as a synergistic response between the endo- and exo-glucanases. The results obtained show that the endo-glucanase activity increased during the whole time of growth, independent of the cellulose concentrations (Fig. 7), that the highest activity was obtained after different days of growth. The more cellulose was used, the longer the time required to get the optimal activity of endoglucanase. Similar results were obtained by testing the activity of the aryl-beta-glucosidase. Quite different results were obtained

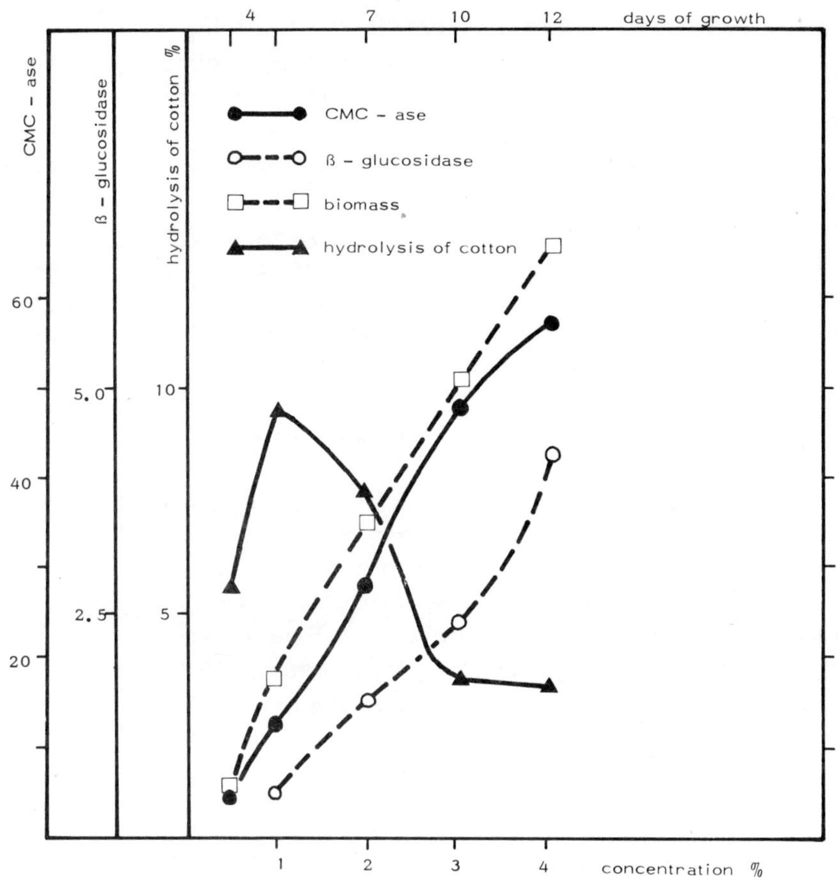

Fig. 7. Effect of different cellulose concentrations on the cellulolytic activity.

assaying the synergistic effect of endo- and exo-glucanases on dewaxed cotton. From the figure presented, we can see, that when the concentration of cellulose exceeds 1 % we observed a strong inhibition of the ability to hydrolyse the cotton. When the culture fluids obtained from cultures with higher cellulose concentration were dialyzed against distilled water (in collodion tubes at 4° during 12 hr) the inhibition was eliminated and the hydrolysis of cotton was then at a similar level to that attained by culturing the fungus on 1 % of cellulose. We are now in the course of doing experiments in which we want to estimate the amounts of cellobiose, and especially the amounts of cellobionolactone, as well as the amounts of glucose and gluconolactone. Only these results can explain the mechanism of a specific inhibition of part of the cellulases complex that have been observed. We are now of the opinion - based especially on the results obtained by the Eriksson group (1974, 1975), that the cellobionolactone specifically inhibits the exo-glucanase (cellobiohydrolase) and that this is the reason for the decreased amounts of hydrolytic products from cotton, whereas the beta-glucosidase, being without excess of substrates, was not inhibited, because it is inhibited only by gluconolactone.

REGULATORY EFFECTS OF DEGRADATION PRODUCTS

Irrespective of how we will explain these relationships, we do have conclusive proof for the existence of a mechanism for cellobionolactone synthesis, and also we have enough information about the inhibitory effect of gluconolactone on glucosidases (cellobiases) (Reese & Parish, 1971; Emert, Gum, Lang & Siu, 1974; Sternberg, 1976) to conclude that both kind of lactones can be a very sensitive mechanism for the microorganisms in the regulation of the amounts of utilizable sugars in the environmental medium. As Emert et al. say: "it would be a competitive advantage to a cellulolytic organism to produce a less universally utilized carbon source, such as cellobiose, rather than glucose".

What is done by the transferase enzyme systems, known in other hydrolytic enzymes, in the case of cellulases by lactonases? Experiments with cellobionic acid and gluconic acid have shown, that both these compounds can be very easily assimilated by *Sp. pulverulentum* (Eriksson, Szajer & Hamp, unpublished) which destroy the balance:

$$\text{gluconolactone} \rightleftharpoons \text{gluconic acid}$$
$$\text{cellobionolactone} \rightleftharpoons \text{cellobionic acid.}$$

Decreasing amounts of lactones in the medium break the inhibitory effect, and active forms of enzymes appear in the mixtures i.e. culture media or in natural ecological systems. So, the important component in these mechanisms seem to be not catabolic repression but the stimulation of the synthesis of both kinds of lactonases. It is possible that this mechanism is a feedback mechanism coupled with the catabolic repression. We can say that there exists a double regulating system in the processes of cellulase synthesis:

1. endocellular, catabolic repression, which cause the standstill of cellulase synthesis
2. exocellular, the production of lactones, acting outside the producers and inhibiting the activity of cellulases.

In this situation, experiments dealing with mutation on "cellulolytic" microorganisms to obtain higher yields of cellulolytic activity are connected with the defect of the ability to synthesize lactonases of both kinds - i.e. cellobionolactonase and glucono-lactonase by the cellulolytic mutants. Further studies are necessary to elaborate some techniques which would make it possible to select mutants of this type. Having such mutants it would be possible to obtain higher yields of cellulase synthesis and also start to explain the mechanism of the catabolic repression during the synthesis of extracellular enzymes by different fungi.

DECOMPOSITION OF LIGNIN

Having in view the application technology of cellulases, we have to take into consideration that most of the cellulose waste substrates, which can be used in technology, are mixtures of cellulose, hemicellulose and the important ingredient lignin. In general, it can be said that the relation of cellulose to lignin is about 40 % to 27 % in spruce and pine and lower in birch wood, - near 21 % (Virkola, 1975). During paper production, the amount of lignins can be reduced to about 3 - 15 %. This is important, because wastepaper is apparently a potential raw material for the enzymatic hydrolysis industry. There are well known experiments dealing with the conversion of cellulose from newsprint waste into glucose. This is a problem of reuse of waste paper: for example, the studies of the Natick group (Mandels, Hontz & Nystrom, 1974; Spano, Medeiros & Mandels, 1975) concerning the production of glucose from newsprint, or the experiments of Eriksson & Larson about the fermentation of waste fibers from a newsprint mill (1975). In all these materials lignin is a component which can be utilized by the microorganism or it can act as a toxin for the growing fungus.

Lignin as a chemical compound is a polymer from one or more phenylpropane monomers. The simplest monomers are: a) hydroxy-cinnamyl alcohol, b) coniferyl alcohol, c) sinapyl alcohol. The most common monomer is coniferyl alcohol, which appears to be the major component of most conifer lignins and forms a significant fraction of the lignin of most other species as well (Brown, 1964). According to Forss & Fremmer (1975), conifer lignin can be considered as an irregular polymer composed of repeated units. Each unit comprises ca. 16 guaiacylpropane units and 2 hydroxyphenylpropane units. The phenylpropane units can be connected with carbohydrates, which bonds are very easily hydrolysed. The microorganisms which are applied in technology or in natural conditions have to have additional abilities, namely the genetical information about the synthesis of lignolytic enzymes. It is interesting to note here, that looking through the papers dealing with lignin biodegradation, that there have so far been described only 2 kinds of enzymes: a) phenol-oxidases, which can be extracellular and endocellular and b) the peroxidase. It is possible, that there could also exist a demethylating enzyme (Trojanowski & Leonowicz, 1963). However, there are reports to the contrary; Gierer & Opara (1973) found evidence that two oxidases, laccase and peroxidase, are not involved in lignolytic activity. In this case, it is not surprising that in the Kirk review (1971) we can find a conclusion, that the specific enzymes involved in the decomposition of lignin by different white-rot fungi are not known. The same author stated, that microorganisms do cause limited changes observed usually as a decrease in methoxyl content or a slight loss of weight, because no specific structures resulting from attack have been elucidated, making it too difficult to generalize as to how lignin is decomposed, especially by white-rot fungi. Selin, Sundman & Raiha (1975) also take into consideration that the ability of some microorganisms to polymerize lignosulfonates means detoxification. Results obtained by these authors have shown that lignosulfonates are utilized by white-rot fungi even if other utilizable carbon sources are present. However, cellulose was not included in these studies. The Eriksson group (Hiroi & Eriksson, 1976; Ander & Eriksson, 1975) published their results about the influence of carbohydrates on lignin degradation. It is very interesting to make a short review of these papers because the experiments show in an excellent way the relationships between microbial cellulose degradation and the degradation of lignincompounds. The experiments were carried out using a wild type of Sp. pulverulentum and a cellulaseless mutant. It was shown that when cellulose was present, the wild type could degrade 60 % of the lignin within three weeks (which means a 3 % lignindegradation per day) and in the absence of cellulose only 36 %, which makes a 1.7 % degradation of lignin per day. In the experiments with the cellulase-less mutant, in an analogous period of time, the mutant degraded only 20 % of lignin when cellulose was added to the medium and only 10 % in three weeks when it was cultured without cellulose. In all cases lignin degradation was favoured by the presence of cellulose and the degradation of lignin was stimulated with the increase of the degree of polymerization of the cellulose. The degradation of lignin follows this order: cotton DP 2000 > cotton DP 500 > Walsetz cellulose DP 150 > cellobiose. Afterwards, they have estimated that the higher the molecular weight of lignocompounds the greater was the influence of the presence of cellulose in the medium upon the degradation rate. The relation between the presence of cellulose and the rapidity of lignin degradation was explained by the existence of the cellobiose:quinone oxidoreductase discovered by Westermark & Eriksson (1974). This enzyme complex oxidized cellobiose to cellobionolactone. This oxidoreductase is induced only when the fungus is grown in media containing cellulose and it facilitates lignin degradation by preventing polymerization reactions catalysed by laccase. Quinones, reaction products of the phenol-oxidizing enzyme laccase, are used as hydrogen receptors. The quinones and the phenol radicals are reduced to the corresponding phenols. These phenols, especially catechol moieties may be readily utilized by other enzymes and further metabolized.

Next Ander & Eriksson (1975, 1976) demonstrated

the importance of phenol oxidase activity in lignin degradation, comparing lignin degradation caused by a phenol oxidase-less mutant with a phenol oxidase positive revertant of Sp. pulverulentum. In these experiments the authors demonstrated that after the addition of highly purified laccase to the kraft lignin medium, the phenol oxidase-less mutant could degrade lignin, and what is more interesting, that the addition of laccase increased the production of endo-1, 4-glucanase in the phenol-oxidase-less mutant in the presence of vanillic acid and kraft lignin.

These experiments have greatly advanced the research into, and increased the understanding of, the biodegradation of lignin compounds. The results obtained explained not only the necessity of laccase in lignin degradation in connection with cellobiose:quinone oxidoreductase, but also presented an additional parameter, the activity of the complex cellobiose:quinone oxidoreductase+laccase, as a measurement to characterize the usefulness of new isolates for technological applications, especially in studies for selecting new mutants with a higher capacity to degrade different ligno-cellulose wastes.

REFERENCES

Ander, P. & Eriksson, K.-E. (1975). Influence of carbohydrates on lignin degradation by the white-rot fungus Sporotrichum pulverulentum. Svensk Papperstidning 18-197578, 643.

Ander, P. & Eriksson, K.-E. (1976). The importance of phenoloxidase activity in lignin degradation by the white-rot fungus Sporotrichum pulverulentum. Arch. Microbiol. 109, 1.

Brown, S. A. (1964). Lignin and tanin biosynthesis. p. 361. In Biochem. of Phenolic Compounds. Ed. by Harborne, J. B., Academic Press, London.

Bujak, S. & Szajer, Cz. (1977). Optimal conditions for cellulase synthesis by Fusarium 25/2 isolate. In preparation for Acta Microbiol. Polon. B.

Dixon, G. R. & Pegg, G. F. (1972). Changes in amino content of tomato xylem sap following infection with strain of Verticillium albo-atrum. Ann. Bot. 36, 147.

Emert, G. H., Gum, E. K., Lang, J. A. & Liu, T. H. (1974). Cellulases. Advances in Chemistry, Ser. 136, 79.

Eriksson, K.-E. & Larson, K. (1975). Fermentation of waste mechanical fibers from a newsprint mill by the rot fungus Sporotrichum pulverulentum. Biotechn. Bioeng. 17, 327.

Eriksson, K.-E., Pettersson, B. & Westermark, U. (1974). Oxidation: an important enzyme reaction in fungal degradation of cellulose. FEBS Lett. 49, 282.

Eriksson, K.-E. & Pettersson, B. (1975). Extracellular system utilized by the fungus Sporotrichum pulverulentum for the breakdown of cellulose. I. Separation, purification and physico-chemical characterization of five endo-1,4 beta-glucanases. Eur. J. Biochem. 51, 193.

Almin, K. E., Eriksson, K.-E. & Pettersson, B. (1975) II. Activities of the five endo-1,4 glucanases towards carboxymethyl-cellulose. Eur. J. Biochem. 51, 207.

Forss, K. & Fremer, K.-E. (1975). Structural model for coniferous lignin and its applicability to the description of the acid bisulfite cook. p. 41. In Symposium on Enzymatic Hydrolysis of Cellulose. Ed. by Bailey, M., Enari, T. M. & Linko, M., SITRA, Helsinki.

Foster, D. M. & Weber, D. J. (1969). Free amino acid pool and enzyme in teratoma and habituated tobacco tissue cultures. Phytopathol. 59, 113.

Ghose, T. K., Pathak, A. N. & Bisaria, V. S. (1975). Kinetic and dynamic studies of Trichoderma viride cellulase production. In: Symposium on Enzymatic Hydrolysis of Cellulose. Ed. by Bailey, M., Enari, T. M. & Linko, M., SITRA, Helsinki.

Gierer, J. & Opara, A. E. (1973). Studies on the enzymatic degradation of lignin. Acta chem. Scand. 27, 2909.

Hanks, R. W. & Feldman, A. W. (1966). Quantitative changes in free and protein amino acids in leaves of healthy Phodopholus similis infected and "recovered" grape-fruit seedlings. Phytopathol. 56, 261.

v. Hofsten, B. (1975). Topological effects in enzymatic and microbial degradation of highly ordered polysaccharides. In: Symposium on Enzymatic Hydrolysis of Cellulose. Ed. by Bailey, M., Enari, T. M. & Linko, M., SITRA, Helsinki.

Hiroi, T. & Eriksson, K.-E. (1976). Microbial degradation of lignin. I. Influence of cellulose on the degradation of lignins by the white-rot fungus Pleurotus ostreatus. Svensk Papperstidning 5-197679, 157.
II. Influence of cellulose upon the degradation of calcium lignosulfonate of various molecular sizes by the white-rot fungus Pleurotus ostreatus. Svenks Papperstidning 5-197679, 163.

Kirk, T. K. (1971). Effect of microorganisms on lignin. Ann. Rev. of Phytopathol. 9, 185.

Lyr, H. & Schanel, L. (1964). Über die Cellulase-Bildung von Pomes marginatus. Zeitschr. Allg. Microbiol. 4, 341.

Mandels, M., Hontz, L. & Nystrom, J. (1974). Enzymatic hydrolysis of waste cellulose. Biotechn. Bioeng. 16, 1471.

Mandels, M. & Weber, J. (1969). The production of cellulases. Advances in Chemistry, Ser. 95, 291.

Nomura, M., Hosoda, J. & Youshikawa, H. (1958). Studies on amylase formation by Bacillus subtilis. VI. The mechanism of amylase excretion and cellular structure of Bac. subtilis. J. Biochem. 45, 737.

Nystrom, J. N. & Kornuta, K. A. (1975). Pilot plant investigations of Trichoderma viride cellulase production. p. 181. In: Symposium on Enzymatic Hydrolysis of Cellulose. Ed. by Bailey, M., Enari, T. M. & Linko, K., SITRA, Helsinki.

Podgórska, E. & Szajer, Cz. (1977). The stability of Fusarium sp. cellulases (in preparation).

Pollock, M. R. (1961). The mechanism of liberation of penicilliase from Bacillus subtilis. J. gen. Microbiol. 26, 267.

Pollock, M. R. (1962). Exoenzymes. In: The Bacteria. Vol. IV, Ch. 4, p. 121. Academic Press, London.

Pettersson, L. G. (1975). The mechanism of enzymatic hydrolysis of cellulose by Trichoderma viride. p. 261. In: Symposium of Enzymatic Hydrolysis of Cellulose. Ed. by Bailey, M., Enari, T. M. & Linko, M., SITRA, Helsinki.

Rees, E. T. & Parish, F. W. (1971). Nojirimycin and d-glucano-1, 5 lactone as inhibitors of carbohydrases. Carbohydrate Res. 18, 381.

Selin, J. F., Sundman, V. & Raiha, M. (1975). Utilization and polymerization of lignosulfonates by wood-rotting fungi. Arch. Microbiol. 103, 63.

Spano, L. A., Medeiros, J. & Mandels, M. (1975). Enzymatic hydrolysis of cellulosic wastes to glucose. Pollution Abatement Division. U.S. Army Natick Development Center.

Sternberg, D. (1976). beta-glucosidase of Trichoderma: its biosynthesis and role in saccharification of cellulose. Applied and Environm. Microbiol. 31, 648.

Szajer, Cz. (1975). Influence of amino acids, thiourea and hydroquinone on cellulolytic activity in some Fusarium strains. Acta Microbiol. Polon., B 7, 111.

Targoński, Z. & Szajer, Cz. (1977). Cellulase activity of Fusarium sp. in different culture conditions. Acta Microbiol. Polon., B (in press).

Targoński, Z. (1977). The activity of the oxidizing enzymes of Fusarium sp. isolate (in preparation).

Trojanowski, J. & Leonowicz, A. (1969). The biodeterioration of lignin in fungi. Microbios. 3, 247.

Trojanowski, J. & Leonowicz, A. (1963). Investigation on the degradation of lignin by Pholiota mutabilis. Ann. Univer. Mariae Curie-Sklodowska, Sec. C 18, 441.

Trojanowski, J. & Leonowicz, A. (1965). Exoenzymes in fungi degrading lignin. Acta Microbiol. Polon. 14, 55.

Virkola, N.-E. (1975). Available cellulosic materials. p. 23. In: Symposium of Enzymatic Hydrolysis of Cellulose. Ed. by Bailey, M., Enari, T. M. & Linko, M. SITRA, Helsinki.

Westermark, U. & Eriksson, K.-E. (1974). Cellobiose: quinone oxidoreductase, a new wood-degrading enzyme from white-rot fungi. Acta chem. Scand., B 28, 209.

Westermark, U. & Eriksson, K.-E. (1975). Purification and properties of cellobiose:quinone oxidoreductase from Sporotrichum pulverulentum. Acta chem. Scand., B 29, 419.

Yoshikawa, T., Suzuki, H. & Nisizawa, K. (1974). Biogenesis of multiple cellulase components of Pseudomonas fluorescens var. cellulose. J. Biochem. 75, 531.

THE MICROBES' POTENTIAL IN RESOURCE ECONOMY

W. R. Stanton

INTRODUCTION

During the last few months, I have had, through the kindness of the Austrian Government, and my host Professor Lafferty at the Institute of Biotechnology at Graz, the opportunity to reflect on the impressions gained in my travels in Africa and Far East and to compare the resource economies of these tropical and equatorial countries (I shall refer to the distinction between the two groups later) on the one hand, and the approach to the utilisation of raw materials in Europe on the other. I have also had the opportunity to consider these resources in relation to land area, productivity, reserves, capital (and its human component in the form of trained manpower), total population and its distribution (Stanton, 1977).

Survey Conclusions

From these reflections two conclusions have appeared: Firstly, that there is in many European countries a greater reserve of under-used recurrent raw materials than in many of the countries of Africa and Asia. There are of course exceptions to this generalization.
Secondly, the machines, the techniques and the processes applied to these resources must be chosen in the context of the total economy of the region, or the territory, for which they are to be considered. Specifically, I concluded that there was a great danger in indiscriminate technological transplantation (Harrison, 1976). The corollary of this is that, though these tropical and equatorial territories may be deficient in the formal understanding of microbial technology,

technology transfer is not a one way process.

Need for Intensive Study of the Tropical World

The Science of Collecting. The student today tends to despise the dusty volumes of 19th century knowledge, the efforts of those patient field workers and collectors who had time to walk about and observe the phenomena around them, record these observations with the old fashioned methods such as the plant press, the pen and steel engraving. This attitude of indifference is also being adopted currently by governments. This is evident from the curtailment of funds. More serious is the fact that the student today does not have the opportunity, even if he had the time, to undertake systematic study; the undisturbed environments no longer exist. At last this erosion of the reservoir of genetic diversity, which applies also to microorganisms, is being recognised and action is being taken, but the remedial measures may be too late.

Traditional Technology. In addition, there is an immense bank of knowledge in the form of the techniques and skills of the local people, which needs to be preserved for all time in the same way as we need to reserve samples of the forms of natural vegetation.

The Need to Conserve "Land Races" of Microorganisms. Advances in antibiotic research and fermenter design have led to a false confidence on the part of the microbiologist on his ability to structure microorganisms as he desires. This may be set down in the form of an hypothesis "that there is more to the genetical structure of an organism and a population in relation to its environment than we have hitherto suspected". This was brought about because under "field" conditions, organisms are subjected to selection over the whole spectrum of environmental pressures, in contrast to the situation in the laboratory or the experimental farm where the range is comparatively narrow. We are just beginning to see the virtue of the former type of selection process in the self optimisation procedures which are being adopted in continuous fermentation.

I will return to this again in a moment in relation to the discussion of the role of the microbe as a part of biotechnology. However, the conclusion reached was that, in the advancement of the biological sciences, time must be made available to remedy deficiences in systematic knowledge. For those concerned with this particular aspect of biology the best defence is to have an adequate answer ready for such questions as, "you have collected samples of fifty different species of this fungal/genus. So what?" It does not mean that the collector should avoid using current techniques, such as sampling strategies and machines, to help in the reduction of the physical task. The latter aid includes automated devices for characterization and storage.

There is a need for self discipline in collecting. This is no new problem to science, but it is far from solved. There is an extreme danger in losing everything in the attempt to conserve everything, even with microorganisms.

Eutrophication and Contamination of Conserved Areas

The best way to conserve these organisms is by conserving the tract of jungle, or mountain on which they live. Diversity can, to some extent, be reconstructed. Plant geneticists have already had to resort to this strategy, but reconstruction of a lost genetic framework is at best difficult and may be impossible, since one does not know what one is trying to construct.

Loss of Diversity in Conserved Areas. There is a hidden danger in that, under conditions of eutrophication, in spite of the efforts of the conservationists to exclude man from the small conserved residue of land, the microorganisms will be the first of the organisms to change their character. This loss of diversity is a well known feature of microbial populations subjected to eutrophication. What is needed is a respect for the set of organelles and molecules operating particular biochemical pathways and building modern structures which have successfully stood the test of biological evolution.

A further hypothesis, arising from the previous remarks, is that "The potential for the application of the microbe is not confined to the manipulation of isolates already to be found in collections. It lies also in the diversity of forms not yet selected, but which have passed the test of selection by evolution in the face of the whole gamut of the biota".

Organisms v. Data. The answer to the question posed earlier on the need for collection is that it is impossible to describe perfectly an organism which is imperfectly characterised and, even if characterised, it might be impossible to reconstruct.

ENERGY ALTERNATIVES

Twelve months ago any argument that there was an alternative to the "limitless energy strategy" was lightly dismissed (Marchetti, 1976). Now energy curtailment strategies are incorporated in the policy of the World's highest *per caput* energy consuming nation, the United States. This basic decision has wide implications, amongst which are decisions on the strategy for microbiology.

Current estimates show that about one fifth, at the present rate of use, of the energy of a country such as Canada may be obtained from the recurrent biomass and in this recovery process microbiology will play its part. However, it is envisaged that the strategy for biomass utilisation will be improved from their present low level (Reed, 1975).

Availability of Substrates

The complaint of the economist and the engineer, heard with increasing frequency since the energy crisis, is that it is not practical, or economical, to collect and process many of the recurrent resources which, in terms of their chemical composition, might be useful. These resources remain unused, or burnt if they cause a nuisance. The microbiologists are to some extent to blame for this attitude, because they have been conditioned by the engineers to identify and isolate only those organisms which fit the engineering processes, rather than consider the equipment of the engineer as at the disposal of the industrial microbiologist to be combined with his own set of operators, the microorganisms themselves. There has been a drift away from biology towards the development of more elaborate models based on the batch or continuous fermenter and the chemostat. Any biological systems which are not open to precise mathematical interpretation, such as those which involve complex substrates like sewage and natural ligno-cellulose, in which the supply of light may be essential and in which several microorganisms grow at the same time, are avoided in the model building process. Though, thankfully, we have seen departures from this attitude in the last 12 months.

Irregularity of Dispersion and Seasonality

The geographical fact must be accepted that recurrent resources are irregularly dispersed and are variable in quality and quantity (Stanton, 1977). There is a tendency to give attention, as far as processing is concerned, solely to those exceptional natural products which occur in large quantities, with regular composition in regular supply throughout the year. Examples of the latter are molasses, bagasse and other secondary products of the equatorial plantation crop industries. Here lies the distinction between biomass production under equatorial (regular) and tropical (monsoonal, seasonal) production.

Combining Biomass Production with Microbiological Conversion

At the recent symposium in Rome on agricultural residues it was recognised that the "wastes" were produced at a high direct energy cost and other indirect energy costs such as fertilisers, tillage, transport, insecticides and fungicides also appeared as a cost of waste production. The greatest waste of all was, of course, the wastage of fixed solar energy.

In team with the bioengineer, the microbiologist is able to achieve a highly efficient usage of the <u>total</u> harvested biomass and conversion of the scarce nutrients, nitrogen and phosphate, to useful products.

The term "energy" is used here in its widest sense to include "energy for food production". This is an important fraction of the total demand for energy by a community and has been highlighted recently in a number of papers on the increasing energy cost of modern agriculture. This energy cost is particularly apparent in the total cost of meat production. By contrast the overall energy budget of "microbial digestible biomass" → single cell protein can be shown to be favourable in comparison with that of "animal digestible biomass" → food animal → animal protein.

Methods of Biomass Conversion by the "Dispersed" Application of Microbial Processes

It is the economic difficulty posed by dispersion which can be overcome by the microbiologist by transferring the process to the substrate, rather than attempting to transport the substrate to the process. This approach is in contrast to the method adopted by for example Dupont-Rank and McClennan (1975), in which large scale processing has been deemed to be the only economical method of microbial processing of biomass. Similar conclusions have been reached concerning municipal wastes at the CRE Symposium in Montreux in 1975. A recent review by Spedding (1976) as identified as whole range of agricultural processes in which there has been a trend away from the biological to the industrial approach and he has stressed the need for the reversal of this trend.

Divergence of Opinion on Future Energy Options

As Spedding has pointed out, divorce from the use of biological systems has proved energy expensive. There are those such as Marchetti (1976) who contend this does not matter, because after the year 2.000 the World will be largely energy independent and be adequately supplied with nuclear energy. There are others who would take the converse view. This has been shown in the widespread concern in the numerous articles, for example, in issues of Der Spiegel in January and February 1977. Thus, two major diametrically opposed points of view occur in Germany alone on future trends in energy usage. Similar concern is widely expressed in other countries. (Nature editorial, 1977).

Large Scale Fermentation v. Chemical Synthesis

If one rejects the hypothesis that microorganisms can assist in economising in the use of energy in alternative agriculture, alternatives <u>to</u> agriculture, manufacturing processes, and direct recovery of solar energy then one simultaneously rejects very many of the future roles of the microorganisms. These current arguments for rejection are powerful, particularly forceful examples being those for ethanol, acetic acid, acetone and glycerol, the recovery of which from recurrent biomass is seen to be quite uneconomical <u>by present standards</u>. (Pape, 1976). If energy is limitless, then the present arguments in favour of encouragement of N-fixation in plants, with its attendant problems, no longer applies.

An Energy Economical Society

If one accepts however the concept of future societies being economical in the use of energy, one returns to the question of the choice of microorganisms and microbial systems mentioned earlier in this paper. In accepting the role of microorganisms in energy economical processes, one may assume that the process aspect of microbiology, biotechnology, will not necessarily be <u>applied</u> to the whole range of potential microbial processes, at least in their present form. The options here will be considered later.

THE ORGANISMIC BASE

Need for Continuing Collection of Microorganisms

As mentioned previously, there is still a need to collect and study microorganisms. This need has been recognised by the UNEP/UNESCO/ICRO Panel for Applied Microbiology in the form of the MIRCENS (Microbial Resource Centres; UNEP/UNESCO/ICRO) project, in which regional centres, in the developing world, are being fostered in relation to major culture collections and microbial-diversity-study centres in the developed world.

Diversity

It is still a matter of speculation as to how great is the diversity of living organisms, especially microbes. It may be assumed to be many times larger in any one habitat than the diversity of higher plants, since each species of higher plant provides niches for many kinds of small organism. For instance, Mellanby (1977) points our, in an article on the need for protection of habitats, that a single species of oak tree (*Quercus* sp.) may provide the home for some 300 different kinds of insect. In Southeast Asia, there are at least ten times as many kinds of oak, as in Austria for example, and yet the oak is regarded as a temperate zone species. Even if the diversity per species of oak is not sustained, the implications for the microbial diversity are that it is enourmous, as each insect species is itself a number of microbial habitats.

Bringing New Groups of Organisms into the Service of Man

Of this total diversity of microorganisms remarkably few have been isolated, identified and domesticated so far. Effort has been concentrated on specific families such as the Lactobacillaceae and Saccharomycetaceae.

Unlike higher plants and animals, the process of selection, discounting that applied to modern industrial microorganisms, has been largely indirect, either by means of the substrate, or the process. Our "non-microbiologically minded" food-technology colleagues have inflicted on the public some unpleasant surprises in the course of development of foods with higher water activity and the lowering of anthocyanin content (bitterness factors) to mention two examples. The latter development has been a contributor to the increased incidence of mycotoxins and the former gives rise to many bacterial hazards when the foods are exposed to normal transport, shop-shelf and household conditions of storage.

Rejection of Organisms Difficult to Culture. This does not

mean that unexploited groups of microorganisms have no virtues. Their absence may be purely a cultural or geographical barrier, or they may have been unfairly labelled as "difficult". Such has been the stigma attached to the basidiomycetes (Hungate, in Schlegel, 1976) and, until recently, the higher plant and animal cell cultures were viewed in a similar light. It is noteworthy that clones of the latter have become selected specifically for growth in culture and the stigma attached to the culture of Basidiomycetes has been based on cultural problems associated with a limited range of mushroom-producing forms. The statement is not true of all basidiomycetes, including equatorial white-rot species, as recent work has shown and as has been reported elsewhere at this Symposium.

Unexploited Groups of Microorganisms. Using the organisms available from the plant kingdom, man's efforts at food production have largely been concerned with improving strains of plants which had been identified and encouraged, largely because of their weed-like growth, as "food plants". The present array of microorganisms available for food processing springs from the collection of organisms associated with the post-harvest techniques applied to these food plants and to the directed spoilage of animal products.

The Neolithic Base for Diversity

The modern agriculturalist may justly be criticised for this orthogenetic approach to the advancement of agriculture and the biotechnologist should similarly be criticised for failure to think in terms of bringing engineering, biochemical and microbiological techniques together as elements of novel art, as distinct from mere refinement of ancient art. The microbial exploitation of mineral oil is a response of society to the failure of agriculture. This error in the current philosophy of agriculture is at last being realised and questions are being asked such as, "Where are the recurrent resource energy plantations?", for which agriculture should have seen the need many years ago, but the remarks of the few, who predicted the need, were rejected. In a similar manner there has been ill-concealed doubt over the rationale, in the spectrum of research, of the efforts of those microbiologists who have postulated the use of organisms such as the hydrogen bacteria and the purple bacteria. The proposals of algologists have received similar treatment. This has come about to some extent by failure to recognise the time-scale of invention, a point to which I will refer in the conclusion.

To the biotechnologists such strange groups as liverworts, slime moulds, mycorrhiza, ferns and mosses are unknown quantities, yet the first are interesting "film reactors", among the second are wood-attacking organisms favoured as grazing ground by insects, the third group performs impossible feats of nutrient transfer and the fourth has an impressive capacity to grow in low light in pure stands and survive depredation, without any apparent armature.

It is encouraging to see that marine algae are at last receiving due attention (Roels *et al.*, 1976) and it is predicted that the present practice of piping sewage to the *bottom* of the sea will be changed to *floating* it gently on the top in the floating impoundments developed for containing oil pollution. This will bring the nutrients and sunlight together, rather than separate them by a gulf of low nutrient, low temperature and sunlight sea. Keeping the pollution where it may be seen and treated will reduce the health hazard rather than conceal it.

Having made the best possible use of the wild material, what of the selection therefrom?

Trends in the Selection and Breeding of Future Microorganisms

A general trend will, it is suggested, be towards a concept of greater "robustness" in the microorganisms of the future. We have already seen the start of this selection process in terms of the self-optimisation processes applied to large-scale biomass production, but the technique will need to be applied to the algae, the nitrogen fixing organisms, the phosphorus releasing organisms of the rhizosphere, the methane and hydrogen producing organisms, and the lignocellulose utilising organisms.

The need for genetic stability has been a central concept in breeding industrial microorganisms, but one of the great divides in microbiology is between the genetic requirements in non-aseptic (septic) and sterile cultures. In terms of scale, many of the goals predicted 4 - 5 years ago (Hedén, 1974) have already been reached, but the current request is for equivalence in the control process (computer) and "organismic" reliability. Is this level of reliability attainable? If we study the fate of other developments in technology such as atomic power stations and giant tankers doubt is justified, but in the strain-self-optimising fermentation systems, such as those applied to yeasts on alkanes, bacteria on methanol and fungi on starch, the future seems brighter. The optimisation strategy is well documented and will not therefore be elaborated here.

Selection for biomass production and harvestability in algae is as yet in its infancy and this appears to be an area for development in the near future. Much has been written about the possible transfer of NIF genes between organisms, particularly blue-green algae in rice paddies for example, but a pre-requisite appears to be the need for proven ecological success in the receiving organisms.

Robustness

The trend towards robustness has already been accepted by the breeders of the higher plants, especially in relation to resistance to wheat and other cereal rusts and the blight of potatoes to mention crop plants important to Europe. Considered carefully, the choice of the presently cultured algal genera such as Chlorella, Scenedesmus and Spirulina has been determined on the basis of their robustness. Within a pre-determined range of ecological conditions these organisms will dominate the niche.

Conclusions About Microorganisms Themselves

For the future role of microorganisms two conclusions appear. Fristly, an increased range of microorganisms, at the level of the higher taxa, will be employed. Secondly, more care will be taken to provide stable, genetically self-repairing, robust organisms capable of surviving in competition with others. This will permit a substantial reduction in capital cost in terms of specifying substrate sterility and allow greater use of the technique discussed in section 5 of "bringing the organism and process to the substrate". Just as the computer has been miniaturised to a household apparatus, so will microorganisms retain and extend their original role as domestic and small-scale adjuncts to providing materials useful to man.

THE SUBSTRATES

Return to Use of Recurrent Resources

This is not the place to decry the use of non-recurrent oil and gas resources as substrates for microorganisms. They have been shown to be a valuable aid in bridging a gap in the immediate future requirement for food and this is an overriding economic consideration. Current attention to methane and methanol is philosophically interesting in that these are both non-recurrent and recurrent resources, though the immediate emphasis is on the non-recurrent supplies. The public and the political scientists have had their say on the carcinogen danger arising from the exploitation of mineral oil. For their part the SCP technologists have taken the utmost precautions over the safety of their products. Perhaps now the overriding need for food will determine the issue.

Coal, brown coal and peat have a cost advantage over oil and the futurologists have given long forward projections for their availability. Brown coal is of particular interest in that it is 50 % volatile, but has a lower value than coal for many heating processes. It is predicted that it may play a role in future as a microbial substrate.

Subject to the energy being available, hydrogen is of importance and the efforts of the microbiologists in utilising hydrogen have been noteworthy.

The Ligno-Cellulose Paradox

If one sits on well-drained ground in a tropical rainforest one sits on almost bare soil, yet all around is the sound of a steady patter of falling leaves and twigs and the occasional crash of larger branches. The air is moist and still. The temperature varies little.

Without degradation the floor would accumulate 10 cm or more of ligno-cellulosic débris each year. This potential accumulation can be appreciated if one wanders through the beech woods of middle Europe in November, where the annual biomass production is small in comparison with that of the rainforest, but the low temperatures arrest decay until the spring. Where does this débris go so quickly in equatorial conditions?

Wastes, Ligno-Cellulose

Obviously biodegradable wastes cannot supply all the required substrate for an economy and the balance must be made up from the "energy plantations" and non-recurrent sources. However, the role of wastes can be substantial and, in the author's experience, 50 % wastage of the harvested matter, which may itself be only 30 % of the grown bulk, is a common figure of agricultural crops. The wastage for forestry is higher, except one of the most striking agricultural wastes, i.e. pineapple, with an 80 % wastage. It is found however that in the poorer economies, with low productivity, wastage, even of low-grade ligno-cellulose residues, is comparatively small. A conclusion, from the author's survey of circum-saharan Africa, was that the core problem was that of increasing production, not in the conventional manner, but with the production strategy being determined in relation to feasible bio-techniques which might be employed to improve utilization of the total biomass.

Availability of Ligno-Cellulose. A number of authors have pointed to ligno-cellulose as our most abundant recurrent resource, both as a specific energy-plantation product and as an agro-industrial and forestry residue. To supplement these resources may be added peat, as a semi-recurrent resource, and brown coal, as a non-recurrent commodity.

Alternative Conversion Routes. The previously mentioned

biomass is produced wet e.g. water weeds or even in suspension (e.g. algae) and no single strategy is possible to recover the energy from this biomass. Anderson (loc. cit) has suggested that, via the methane fermentation (itself up to 70 % efficient), the useful work-energy recoverable from the methane is 30 %, giving an overall efficiency of about 20 %. However, as reviewed elsewhere by the author (Stanton, 1977), and from the work emanating from the Natick Laboratory, the direct aerobic conversion of the cellulose to food, or glucose, is also a possible microbial conversion pathway, alternative to the methane fermentation.

Chemical Preparation of Cellulose. Cellulose conversion by microorganism is well documented, but a pre-requisite is a physio-chemical step to prepare the feedstock and doubt has been cast on the economics of this process in comparison with total chemical conversion (Powell & Bu'Lock, 1976). Further, because of the dispersion of much of the potential ligno-cellulose raw material, the Trichoderm process appears to be confined to pre-collected pre-treated feedstocks. Much straw residue appears best utilized via the ruminant animal, advantageously with some pre-treatment and recycling. There remains the under-utilized dispersed ligno-cellulose residues. These have a low digestibility (1 %) to the ruminant. Termites and wood-rotting basidiomycetes can digest ligno-cellulose, the later can digest it complately, but to quote Hungate (1976 - see Schlegel, p. 340) "the process is very slow".

Hungate adds that the microorganisms in termites are more effective than most others and argues that exploratory studies for the anaerobic digestion of wood should be continued. It should not be forgotten that amongst the termites and other insects are species which may be consumed directly and this represents a possible pathway for direct ligno-cellulose conversion to human food (Stanton, 1977). The use of anaerobic processes is however questionable because of their relative slowness.

At this juncture it is desirable to introduce the third component required for the advancement of applied microbiology namely that supplied by the biotechnologist/bioengineer.

THE ROLE OF THE BIOTECHNOLOGIST/BIOENGINEER

The Multidisciplinary Team

The economic microbiologist is a member of a multidisciplinary team. It has been demonstrated dramatically in the last 2 years that, with the aid of the bioengineer, biological processes can be speeded up rapidly.

Earlier in this paper the bioengineer was critised for his lack of understanding of the contribution which might be made by the whole microorganism. However, a critism may also be levelled at the microbiologist that he has frequently rejected processes on economic grounds, because they were too slow or too difficult to operate, without fully investigating the art of the biological engineer. The gap between the two sciences is most evident in the approach to the treatment of our most abundant recurrent resource ligno-cellulose.

Bioengineering and Ligno-Cellulose

The science of biotechnology and bioengineering is still in the stage of identifying itself (Rehm, 1976) but the simple definition of Taguchi (1976) of "the field which constructs a bridge between applied microbiology and chemical engineering" is apt in the present context. Much has been said of the difficulty of degrading the ligno-cellulose complex and the argument has been that the points of access to the complex, for enzymes, are limited. This assumption has been followed by adopting a biochemist's approach of making <u>one</u> component, cellulose, more freely available to <u>one</u> enzyme system, the cellulase system. The hyphal system of a white-rot organism is however an elegant engineering system, which can penetrate and degrade sound wood, that is ligno-cellulose, in the face of other compounds restricting biochemical activity.

What is doing the degradation and of what components in what order? In practice the high rate is due to a synergism and succession of microorganisms and insects, but the most dramatic single component is that provided by the white rot fungi themselves. Contrary to general conception, some of these white rot basidiomycetes can grow relatively fast, for fungi, but this growth is naturally limited by various constraints such as speed of supply of nutrients, removal of the products of metabolism, temperature and supply of oxygen. However, it was the study of the floor of a rain-forest, plus a primary investigation of biochemical properties of a fast growing basidiomycete (*Volvariella*), which led us to further thoughts on the selection of microorganisms capable of attacking "cellulose".

Rate of "Cellulose" Breakdown and New Developments. A large number of organisms have been screened for "cellulase activity" in static systems, but less attention has been given to screening for systems of ligno-cellulose degradation and to the question, of whether the bioengineer can help and, if so, how? During the last few years features have emerged of the mutual synergism of organisms effecting the degradation process, the synergism between the lignin and cellulose enzyme systems, the lack of interference/suppression by starch and most recently, in a literature search, the author found that a practical system of feeding wood to cattle had been evolved in the past (Kühlwein, 1963), but reports had lain buried, perhaps because the information was ahead of its time. The study of basidiomycete-rotted wood as a feed has received attention (Zadrazil, 1976). An obvious sequel to this was to add a yeast, or other secondary microorganism, to remove the products of metabolism (Stanton, 1977), a ligno-cellulose equivalent to the SYMBA process. A closer study of the findings of Kühlwein showed the crucial organisms were yeasts able to use the products of lignin and that the property identified by Kühlwein might be exhibited by a range of organisms.

Conclusions on Biotechnological Developments

As with the biodegradation of many raw materials of complex constitution, it is necessary to apply a spectrum, or sequence, of microorganisms to effect breakdown. The organisms adapted to the process, as described by Kühlwein, provide the clue as to the type of system which might be optimised by the bioengineer.

Similar remarks may be made about methane fermentation where the bioengineer has been able to help make dramatic improvements in the process rate of the system by control of feed rate, temperature and mixing. Recent studies on manipulation of temperature are particularly interesting as it has recently been shown that operation at elevated temperatures is advantageous, although for many years the figure of 35°C was deemed optimal. It has been demonstrated in the last few years, that with the aid of the arts of the bioengineer, many biological processes can be speeded up rapidly.

THE SYSTEMS APPROACH

It is the author's view that the screening technique used for selection of microorganisms has not taken into account the modern bioengineering developments in the 3 components: oxygen transfer, substrate transfer and product removal.

Screening with Bioengineering in Mind

Screening has tended to operate on static systems, such as strips of filter paper and agar plates, and the data from such results is not applicable to modern high oxygen transfer rate reactors. The limitations occurring under the above heading have been particularly evident in the handling of the microbiological degradation of ligno-cellulose and methane production.

Transport Phenomena

However, limitation of understanding is also to be found in the management of water and the geomeotry of oxidation ponds. In the latter, as has been frequently admitted, the art has largely been in the hands of the civil engineer, without any interaction with the biochemical engineer, or the microbiologist.

Because of the lack of understanding of the transport mechanisms, fixed substrate and film fermentations have been neglected, but there are signs that a fresh interest is being taken in these systems.

Oxygen Limitation

Fixed substrate fermentations are one of the author's particular interests and this particular technique has been given help in recent years by the study of the transport of materials within mycelium. Previously, bioengineers have been concerned that the inside of strands pellets and films is starved of nutrients and oxygen but there has been increasing evidence that this is not so (Trinci in Pirt, 1975).

Commonly, one suspects that heat transfer in pellets and films is better than that predicted on non-biological models. In addition to the above reasons for rejection, solid substrate systems have been rejected on the grounds that contamination is inevitable. If one thinks about it, many processes have been conducted classically, such as cheese production, in which this situation has not posed a serious problem. One of the main causes of contamination is oscillation in humidity and temperature. To avoid these on a large scale may be difficult, but the situation has been overcome classically by the use of caves for solid substrate fermentation, and, if we are all to be driven underground because of the danger of atomic warfare, at least in the intermediate stages the structures built may be used for microbial fermentation. A great virtue of these solid substrate fermentation is that they are energy economical in terms of water addition and removal, prevision of equipment and requirement for maintenance of sterility.

Thus the future, in terms of microbiological systems, will be a broadening one, from a very highly-instrumented spherical and similar geometry high-transfer-rate reactors to low-cost hardware, but nevertheless economical, systems. The new limit is set by the complexity of the raw materials, or by the upper limit of the rate of input of energy to the cell.

RATE AT WHICH INNOVATION TAKES PLACE

Finally, in this review of the microbes potential in resource economy, I wish to bring attention to recent research on the rate at which new inventions are incorporated into the community. Mention has already been made of the work of Marchetti and others reported, at Göttingen Conference (Schlegel & Barnea, 1976). This rationale applies to inventions from all sciences including microbiology. However, since we are concerned with futurology, attention is drawn to the slow rate at which inventions are actually brought into commercial utilisation.

It is not uncommon for an invention to take 50 years to come into operation. It is normal for the process of development of an invention to take 25 years. This is apparent in applied microbiology, particularly in relation to the employment of antibiotics. A similar time lag prevails in relation to fermentation technology in terms of application of new fermenter designs. We should therefore take heart with our research into novel processes and recognise that investigations of the 1950's will only be coming into operation towards the end of the present decade.

The atom bomb neurosis of the microbiological world has been due to the recognition of the widespread occurrence of mycotoxins. But the analogy should not be carried too far. The toxins can theoretically occur all around us, but in practice they do not and, further, there are for the many organisms engaged in synthesising mycotoxins an equal number engaged in degrading them. Even when these poisons have been synthesised there are many barriers to their being specific hazards to human beings and perhaps we should invoke the statistical arguments of the nuclear physicists as to the extent of the hazard caused by mycotoxins. That is, the probability of the death of one person in a hundred thousand, through mycotoxins, should not be regarded the barrier to the development of processes in which they might theoretically occur, although, of course, this argument does not mean that high risk processes should be advocated.

THE FUTURE ROLE OF THE MICROBIOLOGIST

The projection of the future role of microbes in relation to economic development has been on the basis of the need and resources of humanity in relation to a restricted supply of energy. If the additional basis for the projection is accepted, that we are destined to learn to live with the microbes in the future, one can predict an ever widening role of microbiologists in relation to the contact between man and his environment.

REFERENCES

Anderson, R. (1977). Food and fuel self sufficiency through modern biology. Paris: Unesco in press.

Harrison, P. (1976). Inappropriate technology: case histories in a new OECD report. New Scientist, 71, (1011), 236-237.

Hedén, C. G. (1974). Microbial engineering in the next decade. Sao Paulo, Review of advances in biotechnology and bioengineering. Symposium No. 4, 1003-1016.

Hungate, R. E. (1976). Suitability of methanogenic substrates, health hazards, and potential conservation of plant nutrients. In Schlegel, H. G. & Barnea, J. (eds.) (1976).

Kühlwein, H. (1963). Zentr. Bakteriol. Parasitenk. Abt. II, 116, 294.

MacLennan, D. G. (1975). Single cell protein from starch - a new concept in protein production. Food Technology in Australia, April 1975, 141-147.

Marchetti, C. (1976). On properties and behaviour of energy systems. In Schlegel, H. G. & Barnea, J. (eds.) (1976). UNITAR Symposium.

Nature (Editorial) (1977). Carter and the test of energy. 266, 105-111.

Pape, M. (1976). The competition between microbial and chemical processes for the manufacture of basic chemicals and intermediates. In Schlegel, H. G. & Barnea, J. (eds.). UNITAR Symposium.

Powell, I. & Bu'Lock, J. D. (1976). Cellulosic substrates.

Octagon Papers Three, Manchester Univ. Dept. Extra Mural Studies, pp. 155.

Reed, T. B. (1975). Biomass energy refineries for production of fuel and fertilizer. Applied Polymer Symposium No. 28, 1-9.

Rehm, H. J. (1975). Entwicklungen in der Biotechnologie. Chem. Industry XXVII, 373-376.

Roels, O. A., Laurence, S., Farmer, M. W. & van Hemelryek, L. (1975). Organic production potential of artificial upwelling marine culture. In Schlegel, H. G. & Barnea, J. (eds.). UNITAR Symposium.

Schlegel, H. G. & Barnea, J. (eds.) (1976). Proceedings of the UNITAR Symposium, Göttingen "Microbial Energy Conversion". Georg Thieme. Göttingen: Eric Goltzekg pp. 620.

Stanton, W. R. (1977). Microbial resource utilization. Austro-Russian symposium on physiological and technological aspects of biomass production with microorganisms and plant cell culture, Vienna.

Taguchi, H. (1976). Report on education in biochemical engineering. In: Unesco microbiology training course in Japan, Berlin.

Trinci, A. P. J. (1970). Arch. Microbiol. 73, 353.

Zadrazil, H. (1976). Die Verwertung von Abfällen pflanzlicher Herkunft durch Basidiomyceten für Menschen- und Tierernährung. 4th FEMS Symposium, Vienna, Abstracts.

INDEX OF ORGANISMS

SUBJECT INDEX